William Woolsey Johnson

An Elementary Treatise on the Integral Calculus,

founded on the method of rates or fluxions

William Woolsey Johnson

An Elementary Treatise on the Integral Calculus,
founded on the method of rates or fluxions

ISBN/EAN: 9783337308544

Printed in Europe, USA, Canada, Australia, Japan

Cover: Foto ©berggeist007 / pixelio.de

AN

ELEMENTARY TREATISE

ON THE

INTEGRAL CALCULUS

FOUNDED ON THE

METHOD OF RATES OR FLUXIONS

BY

WILLIAM WOOLSEY JOHNSON

PROFESSOR OF MATHEMATICS AT THE UNITED STATES NAVAL ACADEMY
ANNAPOLIS MARYLAND

NEW YORK:
JOHN WILEY AND SONS,
53 EAST TENTH STREET,
1892.

PREFACE.

THIS work, as at present issued, is designed as a shorter course in the Integral Calculus, to accompany the abridged edition of the treatise on the Differential Calculus, by Professor J. Minot Rice and the writer. It is intended hereafter to publish a volume commensurate with the full edition of the work above mentioned, of which the present shall form a part, but which shall contain a fuller treatment of many of the subjects here treated, including Definite Integrals, and the Mechanical Applications of the Calculus, as well as Elliptic Integrals, Differential Equations, and the subjects of Probabilities and Averages. The conception of Rates has been employed as the foundation of the definitions, and of the whole subject of the integration of known functions. The connection between integration, as thus defined, and the process of summation, is established in Section VII. Both of these views of an integral—namely, as a quantity generated at a given rate, and as the limit of a sum—have been freely used in expressing geometrical and physical quantities in the integral form.

The treatises of Bertrand, Frenet, Gregory, Todhunter, and Williamson, have been freely consulted. My thanks are due to Professor Rice for very many valuable suggestions in the course of the work, and for performing much the larger share of the work of revising the proof-sheets.

W. W. J.

U. S. Naval Academy, *July*, 1881.

CONTENTS.

CHAPTER I.

Elementary Methods of Integration.

I.

II.

III.

CHAPTER II.

Methods of Integration—Continued.

IV.

V.

VI.

CHAPTER IV.

Mechanical Applications.

XIV.

XV.

THE

INTEGRAL CALCULUS.

CHAPTER I.

Elementary Methods of Integration.

I.

Integrals.

1. In an important class of problems, the required quantities are magnitudes generated in given intervals of time with rates which are either given in terms of the time t, or are readily expressed in terms of the assumed rate of some other independent variable.

For example, the velocity of a freely falling body is known to be expressed by the equation

$$v = gt, \quad . \quad . \quad . \quad . \quad . \quad . \quad . \quad . \quad (1)$$

in which t is the number of seconds which have elapsed since the instant of rest, and g is a constant which has been determined experimentally. If s denotes the distance of the body at the time t, from a fixed origin taken on the line of motion, v is the rate of s; that is,

$$v = \frac{ds}{dt};$$

hence equation (1) is equivalent to

$$ds = gt\,dt, \qquad (2)$$

which expresses the differential of s in terms of t and dt. Now it is obvious that $\frac{1}{2}gt^2$ is a function of t having a differential equal to the value of ds in equation (2); and, moreover, since two functions which have the same differential (and hence the same rate) can differ only by a constant, the most general expression for s is

$$s = \tfrac{1}{2}gt^2 + C, \qquad (3)$$

in which C denotes an undetermined constant.

2. A variable thus determined from its rate or differential is called an *integral*, and is denoted by prefixing to the given differential expression the symbol $\int$, which is called the integral sign.* Thus, from equation (2) we have

$$s = \int gt\,dt,$$

which therefore expresses that s is a variable whose differential is $gtdt$; and we have shown that

$$\int gt\,dt = \tfrac{1}{2}gt^2 + C.$$

The constant C is called the *constant of integration;* its occurrence in equation (3) is explained by the fact that we have not determined the origin from which s is to be measured.

* The origin of this symbol, which is a modification of the long s, will be explained hereafter. See Art. 100.

If we take this origin at the point occupied by the body when at rest, we shall have $s = 0$ when $t = 0$, and therefore from equation (3) $C = 0$; whence the equation becomes $s = \frac{1}{2}gt^2$.

The Differential of a Curvilinear Area.

3. The area included between a curve, whose equation is given, the axis of x and two ordinates affords an instance of the second case mentioned in the first paragraph of Art. 1; namely, that in which the rate of the generated quantity, although not given in terms of t, can be readily expressed by means of the assumed rate of some other independent variable.

Let BPD in Fig. 1 be the curve whose equation is supposed to be given in the form

$$y = f(x).$$

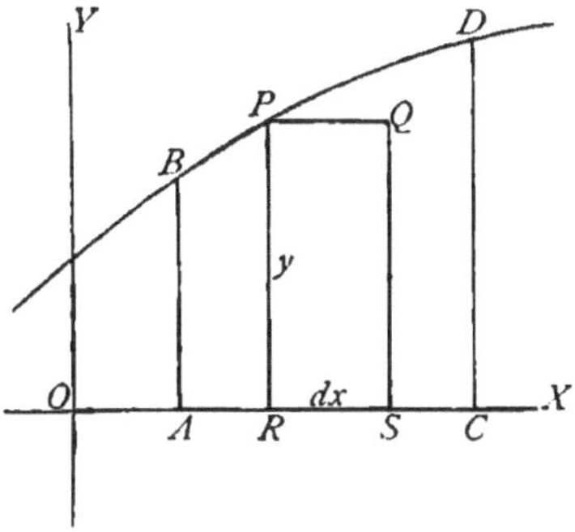

FIG. 1.

Supposing the variable ordinate PR to move from the position AB to the position CD, the required area $ABDC$ is the final value of the variable area $ABPR$, denoted by A, which is generated by the motion of the ordinate. The rate at which the area A is generated can be expressed in terms of the rate of the independent variable x. The required and the assumed rates are denoted, respectively, by $\frac{dA}{dt}$ and $\frac{dx}{dt}$; and, to express the former in terms of the latter, it is necessary to express dA in terms of dx. Since x is an independent variable, we may assume dx to be constant; the rate at which A is generated is then a variable rate, because PR or y is of variable length, while moving at a constant rate along the axis of x. Now dA is the increment which A would receive in the time

dt, were the rate of A to become constant (see Diff. Calc., Art. 17). If, now, at the instant when the ordinate passes the position PR in the figure, its length should become constant, the rate of the area would become constant, and the increment which would then be received in the time dt, namely, the rectangle $PQSR$, represents dA. Since the base RS of this rectangle is dx, we have

$$dA = ydx = f(x)dx. \quad \quad (1)$$

Hence, by the definition given in Art. 2, A is an integral, and is denoted by

$$A = \int f(x)dx. \quad \quad (2)$$

Definite Integrals.

4. Equation (2) expresses that A is a function of x, whose differential is $f(x)dx$; this function, like that considered in Art. 2, involves an undetermined constant. In fact, the expression $\int f(x)dx$ is manifestly insufficient to represent precisely the area $ABPR$, because OA, *the initial value of* x, is not indicated. The indefinite character of this expression is removed by writing this value as a subscript to the integral sign; thus, denoting the initial value by a, we write

$$A = \int_a f(x)dx, \quad \quad (3)$$

in which the subscript is *that value of* x *for which the integral has the value zero.*

If we denote the *final value* of x (OC in the figure) by b, the area $ABDC$, which is a particular value of A, is denoted by writing this value of x at the top of the integral sign, thus,

$$ABDC = \int_a^b f(x)dx. \quad . \quad . \quad . \quad . \quad . \quad . \quad (4)$$

This last expression is called a *definite integral*, and a and b are called its *limits*. In contradistinction, the expression $\int f(x)dx$ is called an *indefinite integral.*

5. As an application of the general expressions given in the last two articles, let the given curve be the parabola

$$y = x^2.$$

Equation (2) becomes in this case

$$A = \int x^2 dx.$$

Now, since $\frac{1}{3}x^3$ is a function whose differential is x^2dx, this equation gives

$$A = \int x^2 dx = \tfrac{1}{3}x^3 + C, \quad . \quad . \quad . \quad . \quad . \quad . \quad (1)$$

in which C is undetermined.

Now let us suppose the limiting ordinates of the required area to be those corresponding to $x = 1$ and $x = 3$. The variable area of which we require a special value is now represented by $\int_1 x^2dx$, which denotes that value of the indefinite integral which vanishes when $x = 1$. If we put $x = 1$ in the general expression in equation (1), namely $\frac{1}{3}x^3 + C$, we have $\frac{1}{3} + C$; hence if we subtract this quantity from the general expression, we shall have an expression which becomes zero when $x = 1$. We thus obtain

$$A = \int_1 x^2 dx = \tfrac{1}{3}x^3 - \tfrac{1}{3}.$$

Finally, putting, in this expression for the variable area, $x = 3$, we have for the required area

$$\int_1^3 x^2 dx = \tfrac{1}{3}3^3 - \tfrac{1}{3} = 8\tfrac{2}{3}.$$

6. It is evident that the definite integral obtained by this process is simply *the difference between the values of the indefinite integral at the upper and lower limits.* This difference may be expressed by attaching the limits to the symbol] affixed to the value of the indefinite integral. Thus the process given in the preceding article is written thus,

$$\int_1^3 x^2 dx = \tfrac{1}{3}x^3 + C\Big]_1^3 = 9 - \tfrac{1}{3} = 8\tfrac{2}{3}.$$

The essential part of this process is the determination of the indefinite integral or function whose differential is equal to the given expression. This is called the *integration* of the given differential expression.

Elementary Theorems.

7. *A constant factor may be transferred from one side of the integral sign to the other.* In other words, if m is a constant and u a function of x,

$$\int mu dx = m\int u dx.$$

Since each member of this equation involves an arbitrary constant, the equation only implies that the two members have the same differential. The differential of an integral is by definition the quantity under the integral sign. Now the second member is the product of a constant by a variable factor; hence its differential is $m\,d\left[\int u\,dx\right]$, that is, $m\,u\,dx$, which is also the differential of the first member.

8. This theorem is useful not only in removing constant factors from under the integral sign, but also in introducing such factors when desired. Thus, given the integral

$$\int x^n\,dx;$$

recollecting that

$$d(x^{n+1}) = (n+1)x^n\,dx,$$

we introduce the constant factor $n+1$ under the integral sign; thus,

$$\int x^n\,dx = \frac{1}{n+1}\int (n+1)x^n\,dx = \frac{1}{n+1}x^{n+1} + C.$$

9. *If a differential expression be separated into parts, its integral is the sum of the integrals of the several parts.* That is, if u, v, w, $\cdots$ are functions of x,

$$\int (u+v+w+\cdots)dx = \int u\,dx + \int v\,dx + \int w\,dx + \cdots$$

For, since the differential of a sum is the sum of the differentials of the several parts, the differential of the second member is identical with that of the first member, and each member involves an arbitrary constant

Thus, for example,

$$\int (2-\sqrt{x})\,dx = \int 2dx - \int x^{\frac{1}{2}}dx = 2x - \tfrac{2}{3}x^{\frac{3}{2}} + C,$$

the last term being integrated by means of the formula deduced in Art. 8.

Fundamental Integrals.

10. The integrals whose values are given below are called the *fundamental integrals.* The constants of integration are generally omitted for convenience.

Formula (*a*) is given in two forms, the first of which is derived in Art. 8, while the second is simply the result of putting $n = -m$. It is to be noticed that this formula gives an indeterminate result when $n = -1$; but in this case, formula (*b*) may be employed.*

The remaining formulas are derived directly from the formulas for differentiation; except that (*j'*), (*k'*), (*l'*), and (*m'*) are derived from (*j*), (*k*), (*l*), and (*m*) by substituting $\frac{x}{a}$ for x.

$$\int x^n dx = \frac{x^{n+1}}{n+1} \qquad \int \frac{dx}{x^m} = -\frac{1}{(m-1)\,x^{m-1}} \quad . \quad . \quad . \quad (a)$$

$$\int \frac{dx}{x} = \log(\pm x)\dagger. \quad . \quad . \quad . \quad . \quad . \quad . \quad . \quad . \quad . \quad (b)$$

$$\int a^x dx = \frac{a^x}{\log a} \qquad \int \varepsilon^x dx = \varepsilon^x \quad . \quad . \quad . \quad . \quad . \quad (c)$$

$$\int \cos\theta\, d\theta = \sin\theta \quad . \quad . \quad . \quad . \quad . \quad . \quad . \quad . \quad . \quad (d)$$

$$\int \sin\theta\, d\theta = -\cos\theta \quad . \quad . \quad . \quad . \quad . \quad . \quad . \quad . \quad . \quad (e)$$

---

* Applying formula (*a*) to the definite integral $\int_a^b x^n dx$, we have

$$\int_a^b x^n dx = \frac{b^{n+1} - a^{n+1}}{n+1},$$

which takes the form $\frac{0}{0}$ when $n = -1$; but, evaluating in the usual manner,

$$\left.\frac{b^{n+1} - a^{n+1}}{n+1}\right]_{n=-1} = \left.\frac{b^{n+1}\log b - a^{n+1}\log a}{1}\right]_{n=-1} = \log b - \log a;$$

a result identical with that obtained by employing formula (*b*).

† That sign is to be employed which makes the logarithm real. See Diff. Calc., Art. 43.

$$\int \frac{d\theta}{\cos^2\theta} = \int \sec^2\theta \, d\theta = \tan\theta \quad . \quad . \quad . \quad . \quad . \quad . \quad . \quad . \quad (f)$$

$$\int \frac{d\theta}{\sin^2\theta} = \int \operatorname{cosec}^2\theta \, d\theta = -\cot\theta \quad . \quad . \quad . \quad . \quad . \quad . \quad . \quad (g)$$

$$\int \frac{\sin\theta \, d\theta}{\cos^2\theta} = \int \sec\theta \tan\theta \, d\theta = \sec\theta \quad . \quad . \quad . \quad . \quad . \quad . \quad . \quad (h)$$

$$\int \frac{\cos\theta \, d\theta}{\sin^2\theta} = \int \operatorname{cosec}\theta \cot\theta \, d\theta = -\operatorname{cosec}\theta \quad . \quad . \quad . \quad . \quad . \quad (i)$$

$$\int \frac{dx}{\sqrt{(1 - x^2)}} = \sin^{-1} x + C = -\cos^{-1} x + C' \quad . \quad . \quad . \quad (j)$$

$$\int \frac{dx}{\sqrt{(a^2 - x^2)}} = \sin^{-1} \frac{x}{a} + C = -\cos^{-1} \frac{x}{a} + C' \quad . \quad . \quad . \quad (j')$$

$$\int \frac{dx}{1 + x^2} = \tan^{-1} x + C = -\cot^{-1} x + C' \quad . \quad . \quad . \quad . \quad (k)$$

$$\int \frac{dx}{a^2 + x^2} = \frac{1}{a} \tan^{-1} \frac{x}{a} + C = -\frac{1}{a} \cot^{-1} \frac{x}{a} + C' \quad . \quad . \quad (k')$$

$$\int \frac{dx}{x\sqrt{(x^2 - 1)}} = \sec^{-1} x + C = -\operatorname{cosec}^{-1} x + C' \quad . \quad . \quad . \quad (l)$$

$$\int \frac{dx}{x\sqrt{(x^2 - a^2)}} = \frac{1}{a} \sec^{-1} \frac{x}{a} + C = -\frac{1}{a} \operatorname{cosec}^{-1} \frac{x}{a} + C' \quad . \quad (l')$$

$$\int \frac{dx}{\sqrt{(2x - x^2)}} = \operatorname{vers}^{-1} x \quad . \quad . \quad . \quad . \quad . \quad . \quad . \quad . \quad . \quad . \quad . \quad (m)$$

$$\int \frac{dx}{\sqrt{(2ax - x^2)}} = \operatorname{vers}^{-1} \frac{x}{a} \quad . \quad . \quad . \quad . \quad . \quad . \quad . \quad . \quad . \quad . \quad (m')$$

Examples I.

Find the values of the following integrals :

1. $\int \frac{dx}{\sqrt{x}}$, $\qquad 2\sqrt{x}$.

2. $\int \frac{dx}{x^2}$, $\qquad -\frac{1}{x}$.

3. $\int_1 \frac{dx}{x^{\frac{3}{2}}}$, $\qquad 2 - \frac{2}{\sqrt{x}}$.

4. $\int_1^\infty \frac{dx}{x^5}$, $\qquad \frac{1}{4}$.

5. $\int_0 \sqrt{x}dx$, $\qquad \frac{2}{3}x^{\frac{3}{2}}$.

6. $\int_1 (x-1)^2 dx$, $\qquad \frac{x^3}{3} - x^2 + x - \frac{1}{3}$.

7. $\int_0^{\frac{a}{b}} (a - bx)^2 dx$, $\qquad a^2x - abx + \frac{b^2x^3}{3}\Big]_0^{\frac{a}{b}} = \frac{a^3}{3b}$.

8. $\int_{-a}^{a} (a + x)^3 dx$, $\qquad a^3x + \frac{3a^2x^2}{2} + ax^3 + \frac{x^4}{4}\Big]_{-a}^{a} = 4a^4$.

9. $\int_1^{a^2} \frac{dx}{x}$, $\qquad 2 \log a$.

10. $\int_{-1}^{-2} \frac{dx}{x}$, $\qquad \log(-x)\Big]_{-1}^{-2} = \log 2$.

11. $\int_a^{4a} \frac{(a+x)^2}{\sqrt{x}}\, dx,$ $\qquad 2\sqrt{x}(a^2 + \tfrac{2}{3}ax + \tfrac{1}{5}x^2)\Big]_a^{4a} = 23\tfrac{11}{15} \cdot a^{\frac{5}{2}}.$

12. $\int_0^{y} \varepsilon^x\, dx,$ $\qquad \varepsilon^y - 1.$

13. $\int_0 \sin\theta\, d\theta,$ $\qquad 1 - \cos\theta.$

14. $\int_0^{\pi} \cos x\, dx,$ $\qquad \sin x\Big]_0^{\pi} = 0.$

15. $\int_0^{\frac{1}{4}\pi} \frac{d\theta}{\cos^2\theta},$ $\qquad \tan\theta\Big]_0^{\frac{1}{4}\pi} = 1.$

16. $\int_0^{\frac{1}{2}a} \frac{dx}{\sqrt{(a^2 - x^2)}},$ $\qquad \sin^{-1}\frac{x}{a}\Big]_0^{\frac{1}{2}a} = \frac{\pi}{6}.$

17. $\int_{-\infty}^{\infty} \frac{dx}{a^2 + x^2},$ $\qquad \frac{\pi}{a}.$

18. $\int_1^{\infty} \frac{dx}{x\sqrt{(x^2 - 1)}}$ $\qquad \frac{\pi}{2}.$

19. If a body is projected vertically upward, its velocity after t units of time is expressed by

$$v = a - gt,$$

a denoting the initial velocity; find the space s_1 described in the time t_1 and the greatest height to which the body will rise.

$$s_1 = \int_0^{t_1} v\, dt = at_1 - \tfrac{1}{2}gt_1^2,$$

$$\text{when } v = 0,\ t = \frac{a}{g},\ s = \frac{a^2}{2g}.$$

20. If the velocity of a pendulum is expressed by

$$v = a \cos \frac{\pi t}{2\tau},$$

the position corresponding to $t = 0$ being taken as origin, find an expression for its position s at the time t, and the extreme positive and negative values of s.

$$s = \frac{2\tau a}{\pi} \sin \frac{\pi t}{2\tau},$$

$$s = \pm \frac{2\tau a}{\pi} \text{ when } t = \tau, 3\tau, 5\tau, \text{ etc.}$$

21. Find the area included between the axis of x and a branch of the curve

$$y = \sin x. \qquad 2.$$

22. Show that the area between the axis of x, the parabola

$$y^2 = 4ax,$$

and any ordinate is two thirds of the rectangle whose sides are the ordinate and the corresponding abscissa.

23. Find (α) the area included by the axes, the curve

$$y = \varepsilon^x,$$

and the ordinate corresponding to $x = 1$, and (β) the whole area between the curve and axes on the left of the axis of y.

$$(\alpha)\ \varepsilon - 1, \ (\beta)\ 1.$$

24. Find the area between the parabola of the nth degree,

$$a^{n-1}y = x^n,$$

and the coordinates of the point (a, a). $\dfrac{a^2}{n+1}$.

25. Show that the area between the axis of x, the rectangular hyperbola

$$xy = 1,$$

the ordinate corresponding to $x = 1$, and any other ordinate is equivalent to the Napierian logarithm of the abscissa of the latter ordinate.

For this reason Napierian logarithms are often called hyperbolic logarithms.

26. Find the whole area between the axes, the curve

$$y^n x^m = a^{n+m},$$

and the ordinate for $x = a$, m and n being positive.

If $n > m$, $\qquad \dfrac{n a^2}{n - m}$;

if $n \leqq m$, $\qquad \infty$.

27. If the ordinate BR of any point B on the circle

$$x^2 + y^2 = a^2$$

be produced so that $BR \cdot RP = a^2$, prove that the whole area between the locus of P and its asymptotes is double the area of the circle.

28. Find the whole area between the axis of x and the curve

$$y(a^2 + x^2) = a^3.$$

πa^2.

29. Find the area between the axis of x and one branch of the *companion to the cycloid*, the equations of which are

$$x = a\psi \qquad y = a(1 - \cos\psi).$$

$2\pi a^2$.

II.

Direct Integration.

11. In any one of the formulas of Art. 10, we may of course substitute for x and dx any function of x and its differential. For instance, if in formula (b) we put $x - a$ in place of x, we have

$$\int \frac{dx}{x-a} = \log(x-a) \qquad \text{or} \qquad \log(a-x),$$

according as x is greater or less than a.

When a given integral is obviously the result of such a substitution in one of the fundamental integrals, or can be made to take this form by the introduction of a constant factor, it is said to be *directly integrable*. Thus, $\int \sin mx\, dx$ is directly integrable by formula (c); for, if in this formula we put mx for θ, we have

$$\int \sin mx \cdot m\, dx = -\cos mx,$$

hence

$$\int \sin mx\, dx = \frac{1}{m}\int \sin mx \cdot m\, dx = -\frac{1}{m}\cos mx.$$

So also in $$\int \sqrt{(a+bx^2)}\, x\, dx,$$

the quantity $x\,dx$ becomes the differential of the binomial $(a + bx^2)$ when we introduce the constant factor $2b$, hence this integral can be converted into the result obtained by putting $(a + bx^2)$ in place of x in $\int \sqrt{x}dx$, which is a case of formula (a). Thus

$$\int \sqrt{(a+bx^2)}\, x\, dx = \frac{1}{2b}\int (a+bx^2)^{\frac{1}{2}}\, 2bx\, dx = \frac{1}{3b}(a+bx^2)^{\frac{3}{2}}.$$

12. A simple algebraic or trigonometric transformation sometimes suffices to render an expression directly integrable, or to separate it into directly integrable parts. Thus, since $-\sin x\,dx$ is the differential of $\cos x$, we have by formula (b)

$$\int \tan x\,dx = \int \frac{\sin x\,dx}{\cos x} = -\log \cos x.$$

So also, by formula (f),

$$\int \tan^2 \theta\,d\theta = \int (\sec^2 \theta - 1)\,d\theta = \tan \theta - \theta;$$

by (e) and (a),

$$\int \sin^3 \theta\,d\theta = \int (1 - \cos^2 \theta) \sin \theta\,d\theta = -\cos \theta + \tfrac{1}{3} \cos^3 \theta;$$

by (j) and (a),

$$\int \sqrt{\left(\frac{1+x}{1-x}\right)}dx = \int \frac{1+x}{\sqrt{(1-x^2)}}dx$$

$$= \int \frac{dx}{\sqrt{(1-x^2)}} - \frac{1}{2}\int (1-x^2)^{-\frac{1}{2}}(-2x\,dx) = \sin^{-1} x - \sqrt{(1-x^2)}.$$

Rational Fractions.

13. When the coefficient of dx in an integral is a fraction whose terms are rational functions of x, the integral may generally be separated into parts directly integrable. If the denominator is of the first degree, we proceed as in the following example.

Given the integral $\displaystyle\int \frac{x^2 - x + 3}{2x + 1}dx$;

by division,

$$\frac{x^2 - x + 3}{2x + 1} = \frac{x}{2} - \frac{3}{4} + \frac{15}{4}\frac{1}{2x + 1},$$

hence

$$\int \frac{x^2 - x + 3}{2x + 1}dx = \frac{1}{2}\int x\,dx - \frac{3}{4}\int dx + \frac{15}{4}\int \frac{dx}{2x + 1}$$

$$= \frac{x^2}{4} - \frac{3x}{4} + \frac{15}{8}\log(2x + 1).$$

When the denominator is of higher degree, it is evident that we may, by division, make the integration depend upon that of a fraction in which the degree of the numerator is lower than that of the denominator by at least a unit. We shall consider therefore fractions of this form only.

Denominators of the Second Degree.

14. If the denominator is of the second degree, it will (after removing a constant, if necessary) either be the square of an expression of the first degree, or else such a square increased or diminished by a constant. As an example of the first case, let us take

$$\int \frac{x + 1}{(x - 1)^2}dx.$$

The fraction may be decomposed thus:

$$\frac{x + 1}{(x - 1)^2} = \frac{x - 1 + 2}{(x - 1)^2} = \frac{1}{x - 1} + \frac{2}{(x - 1)^2};$$

hence

$$\int \frac{x + 1}{(x - 1)^2}dx = \int \frac{dx}{x - 1} + 2\int \frac{dx}{(x - 1)^2}$$

$$= \log(x - 1) - \frac{2}{x - 1}.$$

15. The integral $\displaystyle\int \frac{x + 3}{x^2 + 2x + 6}dx$

affords an example of the second case, for the denominator may be written in the form

$$x^2 + 2x + 6 = (x + 1)^2 + 5.$$

Decomposing the fraction as in the preceding article,

$$\frac{x + 3}{(x + 1)^2 + 5} = \frac{x + 1}{(x + 1)^2 + 5} + \frac{2}{(x + 1)^2 + 5};$$

whence

$$\int \frac{x + 3}{x^2 + 2x + 6} dx = \int \frac{(x + 1)\, dx}{(x + 1)^2 + 5} + 2\int \frac{dx}{(x + 1)^2 + 5}.$$

The first of the integrals in the second member is directly integrable by formula (b), since the differential of the denominator is $2(x + 1)\, dx$, and the second is a case of formula (k'). Therefore

$$\int \frac{x + 3}{x^2 + 2x + 6} dx = \tfrac{1}{2} \log (x^2 + 2x + 6) + \frac{2}{\sqrt{5}} \tan^{-1} \frac{x + 1}{\sqrt{5}}.$$

16. To illustrate the third case, let us take

$$\int \frac{2x + 1}{x^2 - x - 6} dx,$$

in which the denominator is equivalent to $(x - \frac{1}{2})^2 - 6\frac{1}{4}$, and can therefore be resolved into real factors of the first degree. We can then decompose the fraction into fractions having these factors for denominators. Thus, in the present example, assume

$$\frac{2x + 1}{x^2 - x - 6} = \frac{A}{x - 3} + \frac{B}{x + 2}, \quad . \quad . \quad . \quad . \quad . \quad (1)$$

in which A and B are numerical quantities to be determined. Multiplying by $(x - 3)(x + 2)$,

$$2x + 1 = A(x + 2) + B(x - 3). \quad . \quad . \quad . \quad . \quad . \quad (2)$$

Since equation (2) is an algebraic identity, we may in it assign any value we choose to x. Putting $x = 3$, we find

$$7 = 5A, \qquad \text{whence} \qquad A = \tfrac{7}{5},$$

putting $x = -2$,

$$-3 = -5B, \qquad \text{whence} \qquad B = \tfrac{3}{5}.$$

Substituting these values in (1),

$$\frac{2x + 1}{x^2 - x - 6} = \frac{7}{5(x-3)} + \frac{3}{5(x+2)},$$

whence

$$\int \frac{2x+1}{x^2 - x - 6}dx = \tfrac{7}{5}\log(x-3) + \tfrac{3}{5}\log(x+2).$$

17. If the denominator, in a case of the kind last considered, is denoted by $(x - a)(x - b)$, a and b are evidently the roots of the equation formed by putting this denominator equal to zero. The cases considered in Art. 14 and Art. 15 are respectively those in which the roots of this equation are equal, and those in which the roots are imaginary. When the roots are real and unequal, if the numerator does not contain x, the integral can be reduced to the form

$$\int \frac{dx}{(x-a)(x-b)},$$

and by the method given in the preceding article we find

$$\int \frac{dx}{(x-a)(x-b)} = \frac{1}{a-b}\Big[\log(x-a) - \log(x-b)\Big]$$

$$= \frac{1}{a-b}\log\frac{x-a}{x-b}, \quad . \quad . \quad . \quad . \quad . \quad . \quad (A)^*$$

* The formulas of this series are collected together at the end of Chapter II., for convenience of reference. See Art. 101.

in which, when $x < a$, $\log(a - x)$ should be written in place of $\log(x - a)$. [See note on formula (b), Art. 10.]

If $b = -a$, this formula becomes

$$\int \frac{dx}{x^2 - a^2} = \frac{1}{2a} \log \frac{x - a}{x + a} \quad . \quad . \quad . \quad . \quad . \quad (A')$$

Integrals of the special forms given in (A) and (A') may be evaluated by the direct application of these formulas. Thus, given the integral

$$\int \frac{dx}{2x^2 + 3x - 2};$$

if we place the denominator equal to zero, we have the roots $a = \frac{1}{2}$, $b = -2$; whence by formula (A),

$$\int \frac{dx}{2x^2 + 3x - 2} = \frac{1}{2} \int \frac{dx}{(x - \frac{1}{2})(x + 2)} = \frac{1}{2} \cdot \frac{1}{2\frac{1}{2}} \log \frac{x - \frac{1}{2}}{x + 2};$$

or, since $\log(2x - 1)$ differs from $\log(x - \frac{1}{2})$ only by a constant, we may write

$$\int \frac{dx}{2x^2 + 3x - 2} = \frac{1}{5} \log \frac{2x - 1}{x + 2}.$$

Denominators of Higher Degree.

18. When the denominator is of a degree higher than the second, we may in like manner suppose it resolved into factors corresponding to the roots of the equation formed by placing it equal to zero. The fraction (of which we suppose the numerator to be lower in degree than the denominator) may now be decomposed into partial fractions. If the roots are all real and unequal, we assume these partial fractions as in Art. 16; there being one assumed fraction for each factor.

If, however, a pair of imaginary roots occurs, the factor corresponding to the pair is of the form $(x-\alpha)^2+\beta^2$, and the partial fraction must be assumed in the form

$$\frac{Ax+B}{(x-\alpha)^2+\beta^2};$$

for we are only entitled to assume that the numerator of each partial fraction is lower in degree than its denominator (otherwise the given fraction which is the sum of the partial fractions would not have this property).

19. For example, given

$$\int\frac{x+3}{(x^2+1)(x-1)}dx.$$

Assume

$$\frac{x+3}{(x^2+1)(x-1)}=\frac{Ax+B}{x^2+1}+\frac{C}{x-1}, \quad . \quad . \quad . \quad (1)$$

whence

$$x+3=(x-1)(Ax+B)+(x^2+1)C.$$

Putting $x=1$,

$$4=2C, \qquad \text{whence} \qquad C=2;$$

putting $x=0$,

$$3=-B+C, \quad \text{whence} \qquad B=-1.$$

To determine A, any convenient third value may be given to x; for example, if we put $x=-1$, we have

$$2=-2(-A+B)+2C \qquad \therefore \qquad A=-2.$$

Substituting in (1),

$$\frac{x+3}{(x^2+1)(x-1)}=\frac{2}{x-1}-\frac{2x+1}{x^2+1},$$

therefore

$$\int \frac{x+3}{(x^2+1)(x-1)}dx = 2\int\frac{dx}{x-1} - \int\frac{2x\,dx}{x^2+1} - \int\frac{dx}{x^2+1}$$

$$= 2\log(x-1) - \log(x^2+1) - \tan^{-1}x.$$

20. If the denominator admits of factors which are functions of x^2, and the numerator is also a function of x^2, we may with advantage first decompose into fractions having these factors for denominators. Thus, given

$$\int\frac{x^2dx}{x^4-a^4}.$$

Putting y for x^2 in the fraction, we first find

$$\frac{y}{y^2-a^4} = \frac{1}{2(y+a^2)} + \frac{1}{2(y-a^2)},$$

hence

$$\int\frac{x^2dx}{x^4-a^4} = \tfrac{1}{2}\int\frac{dx}{x^2-a^2} + \tfrac{1}{2}\int\frac{dx}{x^2+a^2},$$

therefore [see equation (A'), Art. 17],

$$\int\frac{x^2dx}{x^4-a^4} = \frac{1}{4a}\log\frac{x-a}{x+a} + \frac{1}{2a}\tan^{-1}\frac{x}{a}.$$

This method may sometimes be employed when the numerator is not a function of x^2; thus, since

$$\frac{1}{x^4-a^4} = \frac{1}{2a^2(x^2-a^2)} - \frac{1}{2a^2(x^2+a^2)},$$

we have

$$\frac{x}{x^4-a^4} = \frac{x}{2a^2(x^2-a^2)} - \frac{x}{2a^2(x^2+a^2)},$$

hence

$$\int \frac{x\,dx}{x^4 - a^4} = \frac{1}{4a^2} \log \frac{x^2 - a^2}{x^2 + a^2}.$$

21. The fraction corresponding to a pair of equal roots, that is, to a factor in the denominator of the form $(x - a)^2$, is (see Art. 14) equivalent to a pair of fractions of the form

$$\frac{A}{x - a} + \frac{B}{(x - a)^2},$$

we may, therefore, at once assume the partial fractions in this form. We proceed in like manner when a higher power of a linear factor occurs. For example, given

$$\int \frac{x + 2}{(x - 1)^3 (x + 1)}\,dx;$$

we assume

$$\frac{x + 2}{(x - 1)^3 (x + 1)} = \frac{A}{(x - 1)^3} + \frac{B}{(x - 1)^2} + \frac{C}{x - 1} + \frac{D}{x + 1}.$$

whence

$$x + 2 = [A + B(x - 1) + C(x - 1)^2](x + 1) + D(x - 1)^3. \quad . \quad (1)$$

Putting $x = 1$, we have

$$3 = 2A \qquad \therefore \qquad A = \tfrac{3}{2}.$$

The values of B and C may be determined as follows: if we substitute the value just determined for A, equation (1), is identically satisfied by $x = 1$, hence it may be divided by $x - 1$. We thus obtain

$$-\tfrac{1}{2} = [B + C(x - 1)](x + 1) + D(x - 1)^2 \quad . \quad . \quad (2)$$

in which we may again put $x = 1$, whence $B = -\frac{1}{4}$. In like manner from (2), we obtain

$$\tfrac{1}{4} = C(x+1) + D(x-1),$$

from which $C = \frac{1}{8}$, and $D = -\frac{1}{8}$. Therefore

$$\int \frac{x+2}{(x-1)^3(x+1)}\,dx = \frac{3}{2}\int \frac{dx}{(x-1)^3} - \frac{1}{4}\int \frac{dx}{(x-1)^2} + \frac{1}{8}\int \frac{dx}{x-1} - \frac{1}{8}\int \frac{dx}{x+1}$$

$$= -\frac{3}{4(x-1)^2} + \frac{1}{4(x-1)} + \frac{1}{8}\log\frac{x-1}{x+1}.$$

22. In this example, after obtaining the values of A and D from equation (1) by putting $x = 1$, and $x = -1$, two equations from which B and C might be obtained by elimination could have been derived by giving to x any two other values. Convenient equations for determining B and C may also be obtained by putting $x = 1$ in two equations successively derived by differentiation from the identical equation (1). In the first differentiation we may reject all terms containing $(x-1)^2$; since these terms, and also those derived from them by the second differentiation, will vanish when $x = 1$. Thus, from equation (1), Art. 21, we obtain

$$1 = A + 2Bx + 2C(x^2 - 1) + \text{terms containing } (x-1)^2.$$

Putting $x = 1$, and $A = \frac{3}{2}$, we have $B = -\frac{1}{4}$. Differentiating again and substituting the value of B,

$$0 = -\tfrac{1}{2} + 4Cx + \text{terms containing } (x-1),$$

and, putting $x = 1$ in this last equation, $C = \frac{1}{8}$.

23. When the method of differentiation is applied to a case in which more than one multiple root occurs, it is best to proceed with each root separately. Thus given,

$$\int \frac{x+1}{(x-1)^2(x+2)^2}dx,$$

$$\frac{x+1}{(x-1)^2(x+2)^2} = \frac{A}{(x-1)^2} + \frac{B}{x-1} + \frac{C}{(x+2)^2} + \frac{D}{x+2}$$

whence

$$x+1 = [A + B(x-1)](x+2)^2 + [C + D(x+2)](x-1)^2 \quad . \; . \quad (1)$$

Putting $x = 1$, and $x = -2$, we derive

$$A = \frac{2}{9}, \qquad C = -\frac{1}{9}.$$

Differentiating (1), we have

$$1 = 2A(x+2) + B(x+2)^2 + \text{terms containing } (x-1),$$

whence, putting $x = 1$, and $A = \frac{2}{9}$, we have $B = -\frac{1}{27}$.

Again, differentiating (1), we have

$$1 = 2C(x-1) + D(x-1)^2 + \text{terms containing } (x+2),$$

whence, putting $x = -2$, and $C = -\frac{1}{9}$, we have $D = \frac{1}{27}$.

Therefore

$$\int \frac{x+1}{(x-1)^2(x+2)^2}dx = -\frac{2}{9(x-1)} + \frac{1}{9(x+2)} + \frac{1}{27}\log\frac{x+2}{x-1}.$$

24. Instead of assuming the partial fractions with undetermined numerators, it is sometimes possible to proceed more expeditiously as in the following examples:

Given

$$\int \frac{1}{x^3(1+x^2)}\,dx;$$

putting the numerator in the form $1 + x^2 - x^2$, we have

$$\int \frac{1}{x^3(1+x^2)}\,dx = \int \frac{1+x^2}{x^3(1+x^2)}\,dx - \int \frac{x^2}{x^3(1+x^2)}\,dx$$

$$= \int \frac{dx}{x^3} - \int \frac{1}{x(1+x^2)}\,dx.$$

Treating the last integral in like manner,

$$\int \frac{1}{x^3(1+x^2)}\,dx = \int \frac{dx}{x^3} - \int \frac{dx}{x} + \int \frac{x\,dx}{1+x^2}$$

$$= -\frac{1}{2x^2} - \log x + \tfrac{1}{2}\log(1+x^2) = -\frac{1}{2x^2} + \log\frac{\sqrt{(1+x^2)}}{x}.$$

Again, given

$$\int \frac{1}{x^2(1+x)^2}\,dx;$$

putting the numerator in the form $(1+x)^2 - 2x - x^2$, we have

$$\int \frac{1}{x^2(1+x)^2}\,dx = \int \frac{dx}{x^2} - \int \frac{2+x}{x(1+x)^2}\,dx$$

$$= \int \frac{dx}{x^2} - 2\int \frac{dx}{x(1+x)} + \int \frac{dx}{(1+x)^2}.$$

Hence by equation (A), Art. 17,

$$\int \frac{dx}{x^2(1+x)^2} = -\frac{1}{x} - 2\log\frac{x}{1+x} - \frac{1}{1+x}.$$

Examples II.

1. $\displaystyle\int\frac{dx}{a-x}$, $\qquad -\log(a-x)$.

2. $\displaystyle\int\frac{dx}{(a-x)^2}$, $\qquad \dfrac{1}{a-x}$.

3. $\displaystyle\int\frac{x\,dx}{a^2+x^2}$, $\qquad \dfrac{1}{2}\log(a^2+x^2)$.

4. $\displaystyle\int_0 \sqrt{(a^2-x^2)}\,x\,dx$, $\qquad \dfrac{a^3-(a^2-x^2)^{\frac{3}{2}}}{3}$.

5. $\displaystyle\int_0\frac{x^2\,dx}{a^3-x^3}$, $\qquad \dfrac{1}{3}\log\dfrac{a^3}{a^3-x^3}$.

6. $\displaystyle\int_0\frac{x\,dx}{\sqrt{(a^2-x^2)}}$, $\qquad a-\sqrt{(a^2-x^2)}$.

7. $\displaystyle\int(a^2+3x^2)^3\,x\,dx$, $\qquad \dfrac{(a^2+3x^2)^4}{24}$.

8. $\displaystyle\int_0(a+mx)^2\,dx$, $\qquad \dfrac{(a+mx)^3-a^3}{3m}$.

9. $\displaystyle\int\frac{dx}{\sin^2 2x}$, $\qquad -\dfrac{\cot 2x}{2}$.

10. $\displaystyle\int_0\cos^3 x\sin x\,dx$, $\qquad \dfrac{1-\cos^4 x}{4}$.

11. $\displaystyle\int\frac{\cos\theta\,d\theta}{\sin^3\theta}$, $\qquad -\dfrac{1}{2}\operatorname{cosec}^2\theta$.

12. $\displaystyle\int_0\sec^3 3x\tan 3x\,dx$, $\qquad \dfrac{\sec^3 3x-1}{9}$.

13. $\int a^{mx}dx,$ $\qquad \dfrac{a^{mx}}{m \log a}.$

14. $\int (\varepsilon^x - 1)^3 dx,$ $\qquad \frac{1}{3}\varepsilon^{3x} - \frac{3}{2}\varepsilon^{2x} + 3\varepsilon^x - x.$

15. $\int (1 + 3\sin^2 x)^2 \sin x \cos x\, dx,$ $\qquad \dfrac{(1 + 3\sin^2 x)^3}{18}.$

16. $\int_0^{2a} \dfrac{(a - x)\, dx}{\sqrt{(2ax - x^2)}},$ $\qquad \sqrt{(2ax - x^2)}\Big]_0^{2a} = 0.$

17. $\int_0^{\frac{\pi}{2}} \cos^3 \theta\, d\theta,$ $\qquad \dfrac{2}{3}.$

18. $\int \sec^4 \theta\, d\theta,$ $\qquad \tan\theta + \dfrac{1}{3}\tan^3\theta.$

19. $\int \tan^3 x\, dx,$ $\qquad \dfrac{1}{2}\tan^2 x + \log\cos x.$

20. $\int_0^{\frac{\pi}{4}} \sec^4 x \tan x\, dx,$ $\qquad \dfrac{1}{4}\sec^4 x\Big]_0^{\frac{\pi}{4}} = \dfrac{3}{4}.$

21. $\int \sqrt{\dfrac{a - x}{a + x}}\, dx,$ $\qquad a\sin^{-1}\dfrac{x}{a} + \sqrt{(a^2 - x^2)}.$

22. $\int_{\frac{\pi}{4}}^{\frac{\pi}{2}} \cot^3 \theta\, d\theta,$ $\qquad \dfrac{1 - \log 2}{2}.$

23. $\int \sqrt{\dfrac{2a - x}{x}}\, dx,$ $\qquad \sqrt{(2ax - x^2)} + a\,\mathrm{vers}^{-1}\dfrac{x}{a}.$

24. $\int \sin(\alpha - 2\theta)\, d\theta,$ $\qquad \dfrac{\cos(\alpha - 2\theta)}{2}.$

25. $\int \frac{\cos x \, dx}{a - b \sin x}$, $\quad -\frac{1}{b} \log (a - b \sin x)$.

26. $\int_{\frac{\pi}{6}}^{\frac{\pi}{4}} \frac{dx}{\tan x}$, $\quad \frac{1}{2} \log 2$.

27. $\int_{\frac{\pi}{4}}^{\frac{\pi}{2}} \frac{dx}{\tan x}$, $\quad \frac{1}{2} \log 2$.

28. $\int_{\frac{1}{4}}^{\frac{1}{2}} \frac{dx}{x \log x}$, $\quad \log(-\log x)\Big]_{\frac{1}{4}}^{\frac{1}{2}} = -\log 2$.

29. $\int \frac{dx}{\varepsilon^{x} + \varepsilon^{-x}}$, $\quad \tan^{-1} \varepsilon^{x}$.

30. $\int \frac{x^2 \, dx}{x^6 + 1}$, $\quad \frac{1}{3} \tan^{-1} x^3$.

31. $\int \frac{x \, dx}{\sqrt{(a^4 - x^4)}}$, $\quad \frac{1}{2} \sin^{-1} \frac{x^2}{a^2}$.

32. $\int \frac{dx}{\sqrt{(5 - 3x^2)}}$, $\quad \frac{1}{\sqrt{3}} \sin^{-1} \frac{x\sqrt{3}}{\sqrt{5}}$.

33. $\int \frac{dx}{2 + 5x^2}$, $\quad \frac{1}{\sqrt{10}} \tan^{-1} \frac{x\sqrt{5}}{\sqrt{2}}$.

34. $\int_{1}^{\infty} \frac{dx}{x\sqrt{(2x^2 - 1)}}$, $\quad \frac{\pi}{4}$.

35. $\int_{0}^{1} \frac{dx}{x^2 + x + 1}$, $\quad \frac{2}{\sqrt{3}} \tan^{-1} \frac{2x + 1}{\sqrt{3}}\Big]_{0}^{1} = \frac{\pi}{3\sqrt{3}}$.

36. $\int_0^1 \frac{dx}{\sqrt{(5-4x-x^2)}}$, $\cos^{-1}\frac{2}{3}$.

37. $\int \frac{\sqrt{(x^2-a^2)}}{x}\,dx \left[= \int \frac{x^2-a^2}{x\sqrt{(x^2-a^2)}}\,dx\right]$,

$\sqrt{(x^2-a^2)} - a\sec^{-1}\frac{x}{a}$.

38. $\int_0^{\frac{1}{2}a} \frac{x^2}{a-x}\,dx$, $a^2(\log 2 - \frac{5}{8})$.

39. $\int \frac{4x^2-x+3}{x^2+1}\,dx$, $4x - \frac{1}{2}\log(x^2+1) - \tan^{-1}x$.

40. $\int \frac{x^2+x+1}{x^2-x+1}\,dx$, $x + \log(x^2-x+1) + \frac{2}{\sqrt{3}}\tan^{-1}\frac{2x-1}{\sqrt{3}}$.

41. $\int \frac{x^2-1}{x^2-4}\,dx$, $x + \frac{3}{4}\log\frac{x-2}{x+2}$.

42. $\int \frac{(1+x)^2}{x-x^2}\,dx$, $\log\frac{x}{(1-x)^4} - x$.

43. $\int \frac{(2x+1)^2\,dx}{2x+3}$, $x^2 - x + 2\log(2x+3)$.

44. $\int \frac{2x+3}{(2x+1)^2}\,dx$, $\frac{1}{2}\log(2x+1) - \frac{1}{2x+1}$.

45. $\int \frac{x^2-3x+3}{x^2-3x+2}\,dx$, $x + \log\frac{x-2}{x-1}$.

46. $\int_0^a \frac{dx}{x^2-2ax\cos\alpha+a^2}$,

$\frac{1}{a\sin\alpha}\tan^{-1}\frac{x-a\cos\alpha}{a\sin\alpha}\Big]_0^a = \frac{\pi-\alpha}{2a\sin\alpha}$.

47. $\int \frac{dx}{x^2 - 2ax \sec\alpha + a^2}$, $\frac{1}{2a \tan\alpha} \log \frac{x - a\sec\alpha - a\tan\alpha}{x - a\sec\alpha + a\tan\alpha}$.

48. $\int \frac{dx}{2x^2 - 4x - 7}$, $\frac{\sqrt{2}}{12} \log \frac{2x - 2 - 3\sqrt{2}}{2x - 2 + 3\sqrt{2}}$.

49. $\int \frac{x^2\, dx}{1 - x^6}$, $\frac{1}{6} \log \frac{1 + x^3}{1 - x^3}$.

50. $\int \frac{3x - 1}{x^3 - x^2 - 2x}\, dx$, $\frac{1}{6} \log \frac{x^3 (x - 2)^5}{(x + 1)^8}$.

51. $\int \frac{x\, dx}{(x + 2)(x + 3)^2}$, $2 \log \frac{x + 3}{x + 2} - \frac{3}{x + 3}$.

52. $\int \frac{x\, dx}{x^3 + x^2 + x + 1}$, $\frac{1}{2}\left[\tan^{-1} x + \log \frac{\sqrt{(x^2 + 1)}}{x + 1}\right]$.

53. $\int \frac{x^2 dx}{x^4 + x^2 - 2}$, $\frac{1}{6} \log \frac{x - 1}{x + 1} + \frac{\sqrt{2}}{3} \tan^{-1} \frac{x}{\sqrt{2}}$.

54. $\int \frac{x^2 - x + 2}{x^4 - 5x^2 + 4}\, dx$, $\frac{2}{3} \log \frac{x + 1}{x + 2} + \frac{1}{3} \log \frac{x - 2}{x - 1}$.

55. $\int \frac{dx}{x^3 - x^2 - x + 1}$, $\frac{1}{4} \log \frac{x + 1}{x - 1} - \frac{1}{2(x - 1)}$.

56. $\int \frac{3x + 1}{x^4 - 1}\, dx$, $\log \frac{(x - 1)\sqrt{(x + 1)}}{(x^2 + 1)^{\frac{3}{4}}} - \frac{1}{2} \tan^{-1} x$.

57. $\int \frac{dx}{1 + x^3}$, $\frac{1}{6} \log \frac{(x + 1)^2}{x^2 - x + 1} + \frac{1}{\sqrt{3}} \tan^{-1} \frac{2x - 1}{\sqrt{3}}$.

58. $\int \frac{x^2 dx}{(x - 1)^2 (x^2 + 1)}$, $\frac{1}{2} \log (x - 1) - \frac{1}{4} \log (x^2 + 1) - \frac{1}{2(x - 1)}$.

$$\int \frac{x^2\,dx}{1-x^6} = \int \frac{x^2\,dx}{(1-x^3)(1+x^3)}$$

$$\frac{x^2}{(1-x^3)(1+x^3)} = \frac{Ax^2}{1-x^3} + \frac{Bx^2}{1+x^3}$$

$\therefore A = B = 1/2$

$$\int \frac{x^2\,dx}{1-x^6} = \frac{1}{2}\int \frac{x^2\,dx}{1-x^3} + \frac{1}{2}\int \frac{x^2\,dx}{1+x^3}$$

$$= -\frac{1}{6}\log(1-x^3) + \frac{1}{6}\log(1+x^3)$$

$$= \frac{1}{6}\log\frac{1+x^3}{1-x^3}$$

Note that the undetermined quantities in numerators are coefficients of x^2

Ex. 419. page 30.

59. $\displaystyle\int \frac{dx}{x(1 + x + x^2 + x^3)}$,

$$\log x - \frac{1}{2}\log(1 + x) - \frac{1}{4}\log(1 + x^2) - \frac{1}{2}\tan^{-1} x.$$

60. $\displaystyle\int \frac{x^2 - 1}{x^4 + x^2 + 1}\,dx$, $\displaystyle\frac{1}{2}\log\frac{x^2 - x + 1}{x^2 + x + 1}$.

61. $\displaystyle\int \frac{x^2 + x - 1}{x^3 + x^2 - 6x}\,dx$, $\displaystyle\frac{1}{6}\log x + \frac{1}{2}\log(x - 2) + \frac{1}{3}\log(x + 3)$.

62. $\displaystyle\int \frac{x^2 dx}{x^4 - x^2 - 12}$, $\displaystyle\frac{1}{7}\log\frac{x - 2}{x + 2} + \frac{\sqrt{3}}{7}\tan^{-1}\frac{x}{\sqrt{3}}$.

63. $\displaystyle\int \frac{x^2 dx}{(x^2 - 1)^2}$, $\displaystyle\frac{1}{4}\log\frac{x - 1}{x + 1} - \frac{x}{2(x^2 - 1)}$.

64. $\displaystyle\int \frac{2x^3 - 3a^2}{x^4 - a^4}\,dx$, $\displaystyle\frac{5}{2a}\tan^{-1}\frac{x}{a} - \frac{1}{4a}\log\frac{x - a}{x + a}$.

65. $\displaystyle\int \frac{x\,dx}{x^4 - x^2 - 2}$, $\displaystyle\frac{1}{6}\log\frac{x^2 - 2}{x^2 + 1}$.

66. $\displaystyle\int \frac{dx}{(x^2 + a^2)(x + b)}$, $\displaystyle\frac{1}{b^2 + a^2}\left[\log\frac{x + b}{\sqrt{(x^2 + a^2)}} + \frac{b}{a}\tan^{-1}\frac{x}{a}\right]$.

67. $\displaystyle\int_0^\infty \frac{dx}{(x^2 + a^2)(x^2 + b^2)}$, $\displaystyle\frac{\pi}{2ab(a + b)}$.

68. $\displaystyle\int_a^\infty \frac{a^2 dx}{x^2(a^2 + x^2)}$, $\displaystyle\frac{4 - \pi}{4a}$.

69. $\displaystyle\int \frac{x + 1}{x(1 + x^2)}\,dx$, $\displaystyle\tan^{-1} x + \log\frac{x}{\sqrt{(1 + x^2)}}$.

70. $\displaystyle\int \frac{dx}{x^4(x^2 + 1)}$, $\displaystyle\tan^{-1} x + \frac{1}{x} - \frac{1}{3x^3}$.

71. $\int \frac{dx}{x(1+x)^2}$, $\qquad \log \frac{x}{1+x} + \frac{1}{1+x}$.

72. $\int \frac{dx}{x(a+bx^3)}$, $\qquad \frac{1}{3a} \log \frac{x^3}{a+bx^3}$.

73. $\int \frac{dx}{x^4(a+bx^3)}$, $\qquad -\frac{1}{3ax^3} + \frac{b}{3a^2} \log \frac{a+bx^3}{x^3}$.

74. Find the whole area enclosed by both loops of the curve

$$y^2 = x^2(1-x^2). \qquad \tfrac{4}{3}.$$

75. Find the area enclosed between the asymptote corresponding to $x = a$, and the curve

$$x^2y^2 + a^2x^2 = a^2y^2. \qquad 2a^2.$$

76. Find the whole area enclosed by the curve

$$a^2y^4 = x^4(a^2 - x^2). \qquad \tfrac{8}{5}a^2.$$

77. Find the area enclosed by the catenary

$$y = \frac{c}{2}\left[\varepsilon^{\frac{x}{c}} + \varepsilon^{-\frac{x}{c}}\right],$$

the axes and any ordinate.

$$\frac{c^2}{2}\left[\varepsilon^{\frac{x}{c}} - \varepsilon^{-\frac{x}{c}}\right].$$

78. Find the whole area between the witch

$$xy^2 = 4a^2(2a - x)$$

and its asymptote. *See Ex.* 23.

$4\pi a^2$.

III.

Trigonometric Integrals.

25. The transformation, $\tan^2\theta = \sec^2\theta - 1$, suffices to separate all integrals of the form

$$\int \tan^n \theta \, d\theta, \quad \dots \dots \dots \quad (1)$$

in which n is an integer, into directly integrable parts. Thus, for example,

$$\int \tan^5 \theta \, d\theta = \int \tan^3 \theta \, (\sec^2 \theta - 1) \, d\theta$$

$$= \frac{\tan^4 \theta}{4} - \int \tan^3 \theta \, d\theta.$$

Transforming the last integral in like manner, we have

$$\int \tan^5 \theta \, d\theta = \frac{\tan^4 \theta}{4} - \frac{\tan^2 \theta}{2} + \int \tan \theta \, d\theta;$$

hence (see Art. 12)

$$\int \tan^5 \theta \, d\theta = \frac{\tan^4 \theta}{4} - \frac{\tan^2 \theta}{2} - \log \cos \theta.$$

When the value of n in (1) is even, the value of the final integral will be θ. When n is negative, the integral takes the form

$$\int \cot^n \theta \, d\theta,$$

which may be treated in a similar manner.

26. Integrals of the form

$$\int \sec^n \theta \, d\theta \quad . \quad . \quad . \quad . \quad . \quad . \quad . \quad . \quad . \quad (2)$$

are readily evaluated *when* n *is an even number*, thus

$$\int \sec^6 \theta \, d\theta = \int (\tan^2 + 1)^2 \sec^2 \theta \, d\theta$$

$$= \int \tan^4 \theta \sec^2 \theta \, d\theta + 2 \int \tan^2 \theta \sec^2 \theta \, d\theta + \int \sec^2 \theta \, d\theta$$

$$= \frac{\tan^5 \theta}{5} + \frac{2 \tan^3 \theta}{3} + \tan \theta.$$

If n in expression (2) is odd, the method to be explained in Section VI is required.

Integrals of the form $\int \operatorname{cosec}^n \theta \, d\theta$ are treated in like manner.

Cases in which $\sin^m \theta \cos^n \theta \, d\theta$ *is directly integrable.*

27. If n is a *positive odd number*, an integral of the form

$$\int \sin^m \theta \cos^n \theta \, d\theta \quad . \quad . \quad . \quad . \quad . \quad . \quad . \quad . \quad (3)$$

is directly integrable in terms of $\sin \theta$. Thus,

$$\int \sin^2 \theta \cos^5 \theta \, d\theta = \int \sin^2 \theta \, (1 - \sin^2 \theta)^2 \cos \theta \, d\theta$$

$$= \frac{\sin^3 \theta}{3} - \frac{2 \sin^5 \theta}{5} + \frac{\sin^7 \theta}{7}.$$

This method is evidently applicable even when m is fractional or negative. Thus, putting y for $\sin \theta$,

$$\int \frac{\cos^3\theta}{\sin^{\frac{3}{2}}\theta}\,d\theta = \int \frac{(1-y^2)\,dy}{y^{\frac{3}{2}}} = \int y^{-\frac{3}{2}}\,dy - \int y^{\frac{1}{2}}\,dy;$$

hence

$$\int \frac{\cos^3\theta}{\sin^{\frac{3}{2}}\theta}\,d\theta = -2y^{-\frac{1}{2}} - \frac{2}{3}y^{\frac{3}{2}} = -\frac{2}{3}\cdot\frac{3+\sin^2\theta}{\sqrt{(\sin\theta)}}.$$

When m in expression (3) is a positive odd number, the integral is evaluated in a similar manner.

28. An integral of the form (3) is also directly integrable *when* m + n *is an even negative integer*, in other words, when it can be written in the form

$$\int \frac{\sin^m\theta\,d\theta}{\cos^{m+2q}\theta} = \int \tan^m\theta\,\sec^{2q}\theta\,d\theta,$$

in which q *is positive.*

For example,

$$\int \frac{d\theta}{\sin^{\frac{3}{2}}\theta\,\cos^{\frac{5}{2}}\theta} = \int (\tan\theta)^{-\frac{3}{2}}\sec^4\theta\,d\theta$$

$$= \int (\tan\theta)^{-\frac{3}{2}}(\tan^2\theta + 1)\sec^2\theta\,d\theta;$$

hence

$$\int \frac{d\theta}{\sin^{\frac{3}{2}}\theta\,\cos^{\frac{5}{2}}\theta} = \frac{2}{3}\tan^{\frac{3}{2}}\theta - \frac{2}{\tan^{\frac{1}{2}}\theta}.$$

It may be more convenient to express the integral in terms of $\cot\theta$ and $\operatorname{cosec}\theta$, thus

$$\int \frac{\cos^4\theta\,d\theta}{\sin^8\theta} = \int \cot^4\theta\,(\cot^2\theta + 1)\operatorname{cosec}^2\theta\,d\theta$$

$$= -\frac{\cot^7\theta}{7} - \frac{\cot^5\theta}{5}.$$

Integrals of the forms treated in Art. 25 and Art. 26 are included in the general form (3), Art. 27. Except in the cases already considered, and in the special cases given below, the method of reduction given in Section VI is required in the evaluation of integrals of this form.

The Integrals $\int \sin^2 \theta \, d\theta$, *and* $\int \cos^2 \theta \, d\theta$.

29. These integrals are readily evaluated by means of the transformations

$$\sin^2 \theta = \tfrac{1}{2}(1 - \cos 2\theta), \qquad \text{and} \qquad \cos^2 \theta = \tfrac{1}{2}(1 + \cos 2\theta).$$

Thus

$$\int \sin^2 \theta \, d\theta = \tfrac{1}{2}\int d\theta - \tfrac{1}{2}\int \cos 2\theta \, d\theta = \tfrac{1}{2}\theta - \tfrac{1}{4}\sin 2\theta,$$

or, since $\qquad \sin 2\theta = 2 \sin \theta \cos \theta,$

$$\int \sin^2 \theta \, d\theta = \tfrac{1}{2}(\theta - \sin \theta \cos \theta). \quad . \quad . \quad . \quad . \quad (B)$$

In like manner

$$\int \cos^2 \theta \, d\theta = \tfrac{1}{2}(\theta + \sin \theta \cos \theta). \quad . \quad . \quad . \quad . \quad (C)$$

Since $\sin^2 \theta + \cos^2 \theta = 1$, the sum of these integrals is $\int d\theta$; accordingly we find the sum of their values to be θ.

In the applications of the Integral Calculus, these integrals frequently occur with the limits 0 and $\frac{1}{2}\pi$; from (*B*) and (*C*) we derive

$$\int_0^{\frac{\pi}{2}} \sin^2 \theta \, d\theta = \int_0^{\frac{\pi}{2}} \cos^2 \theta \, d\theta = \tfrac{1}{4}\pi.$$

The Integrals $\int\frac{d\theta}{\sin\theta\cos\theta}$, $\int\frac{d\theta}{\sin\theta}$, *and* $\int\frac{d\theta}{\cos\theta}$.

30. We have

$$\int\frac{d\theta}{\sin\theta\cos\theta}=\int\frac{\sec^2\theta\,d\theta}{\tan\theta}=\log\tan\theta. \quad . \quad . \quad . \quad (D)$$

Again, using the transformation,

$$\sin\theta=2\sin\tfrac{1}{2}\theta\cos\tfrac{1}{2}\theta,$$

we have

$$\int\frac{d\theta}{\sin\theta}=\int\frac{\frac{1}{2}\,d\theta}{\sin\frac{1}{2}\theta\cos\frac{1}{2}\theta}=\int\frac{\sec^2\frac{1}{2}\theta\,\frac{1}{2}d\theta}{\tan\frac{1}{2}\theta};$$

hence

$$\int\frac{d\theta}{\sin\theta}=\log\tan\tfrac{1}{2}\theta. \quad . \quad . \quad . \quad . \quad . \quad . \quad . \quad (E)$$

This integral may also be evaluated thus,

$$\int\frac{d\theta}{\sin\theta}=\int\frac{\sin\theta\,d\theta}{\sin^2\theta}=\int\frac{\sin\theta\,d\theta}{1-\cos^2\theta}.$$

Since $\sin\theta\,d\theta=-d(\cos\theta)$, the value of the last integral is, by formula (A'), Art. 17,

$$\frac{1}{2}\log\frac{1-\cos\theta}{1+\cos\theta}=\log\sqrt{\frac{1-\cos\theta}{1+\cos\theta}};$$

and, multiplying both terms of the fraction by $1-\cos\theta$, we have

$$\int\frac{d\theta}{\sin\theta}=\log\frac{1-\cos\theta}{\sin\theta}. \quad . \quad . \quad . \quad . \quad . \quad (E')$$

31. Since $\cos\theta = \sin(\frac{1}{2}\pi + \theta)$, we derive from formula (E),

$$\int\frac{d\theta}{\cos\theta} = \int\frac{d\theta}{\sin(\frac{1}{2}\pi+\theta)} = \log\tan\left[\frac{\pi}{4}+\frac{\theta}{2}\right]. \quad . \quad . \quad (F)$$

By employing a process similar to that used in deriving formula (E'), we have also

$$\int\frac{d\theta}{\cos\theta} = \log\frac{1+\sin\theta}{\cos\theta}. \quad . \quad . \quad . \quad . \quad . \quad (F')$$

Miscellaneous Trigonometric Integrals.

32. A trigonometric integral may sometimes be reduced, by means of the formulas for trigonometric transformation, to one of the forms integrated in the preceding articles. For example, let us take the integral

$$\int\frac{d\theta}{a\sin\theta + b\cos\theta}.$$

Putting $\qquad a = k\cos\alpha, \qquad b = k\sin\alpha, \quad . \quad . \quad . \quad . \quad (1)$

we have

$$\int\frac{d\theta}{a\sin\theta + b\cos\theta} = \frac{1}{k}\int\frac{d\theta}{\sin(\theta+\alpha)}.$$

Hence by formula (E)

$$\int\frac{d\theta}{a\sin\theta + b\cos\theta} = \frac{1}{k}\log\tan\frac{1}{2}(\theta+\alpha);$$

or, since equations (1) give

$$k = \sqrt{(a^2+b^2)}, \qquad \tan\alpha = \frac{b}{a},$$

$$\int\frac{d\theta}{a\sin\theta + b\cos\theta} = \frac{1}{\sqrt{(a^2+b^2)}}\log\tan\frac{1}{2}\left[\theta + \tan^{-1}\frac{b}{a}\right].$$

33. The expression $\sin m\theta \sin n\theta \, d\theta$ may be integrated by means of the formula

$$\cos (m - n) \theta - \cos (m + n) \theta = 2 \sin m\theta \sin n\theta ;$$

whence

$$\int \sin m\theta \sin n\theta \, d\theta = \frac{\sin (m - n) \theta}{2 (m - n)} - \frac{\sin (m + n) \theta}{2 (m + n)} . \quad . \quad (1)$$

In like manner, from

$$\cos (m - n) \theta + \cos (m + n) \theta = 2 \cos m\theta \cos n\theta,$$

we derive

$$\int \cos m\theta \cos n\theta \, d\theta = \frac{\sin (m - n) \theta}{2 (m - n)} + \frac{\sin (m + n) \theta}{2 (m + n)} . \quad . \quad (2)$$

When $m = n$, the first term of the second member of each of these equations takes an indeterminate form. Evaluating this term, we have

$$\int \sin^2 n\theta \, d\theta = \frac{\theta}{2} - \frac{\sin 2n\theta}{4n}, \quad . \quad . \quad . \quad . \quad . \quad (3)$$

and

$$\int \cos^2 n\theta \, d\theta = \frac{\theta}{2} + \frac{\sin 2n\theta}{4n} . \quad . \quad . \quad . \quad . \quad . \quad (4)$$

Using the limits 0 and π we have, from (1) and (2), *when* m *and* n *are unequal integers,*

$$\int_0^\pi \sin m\theta \sin n\theta \, d\theta = \int_0^\pi \cos m\theta \cos n\theta \, d\theta = 0 ; \quad . \quad . \quad (5)$$

but, when m and n are *equal integers,* we have from (3) and (4)

$$\int_0^\pi \sin^2 n\theta \, d\theta = \int_0^\pi \cos^2 n\theta \, d\theta = \frac{\pi}{2} . \quad . \quad . \quad . \quad . \quad . \quad (6)$$

34. To integrate $\sqrt{(1 + \cos \theta)} \, d\theta$, we use the formula

$$2 \cos^2 \tfrac{1}{2}\theta = 1 + \cos \theta,$$

whence $$\sqrt{(1+\cos\theta)} = \pm\sqrt{2}\cos\tfrac{1}{2}\theta,$$

in which the positive sign is to be taken, provided the value of θ is between 0 and π. Supposing this to be the case, we have

$$\int\sqrt{(1+\cos\theta)}\,d\theta = \sqrt{2}\int\cos\tfrac{1}{2}\theta\,d\theta$$
$$= 2\sqrt{2}\sin\tfrac{1}{2}\theta.$$

For example, we have the definite integral

$$\int_0^{\frac{\pi}{2}}\sqrt{(1+\cos\theta)}\,d\theta = 2\sqrt{2}\sin\frac{\pi}{4} = 2.$$

Integration of $\dfrac{d\theta}{a+b\cos\theta}$.

35. By means of the formulas

$$1 = \cos^2\tfrac{1}{2}\theta + \sin^2\tfrac{1}{2}\theta \quad\text{and}\quad \cos\theta = \cos^2\tfrac{1}{2}\theta - \sin^2\tfrac{1}{2}\theta,$$

we have

$$\int\frac{d\theta}{a+b\cos\theta} = \int\frac{d\theta}{(a+b)\cos^2\frac{1}{2}\theta + (a-b)\sin^2\frac{1}{2}\theta}.$$

Multiplying numerator and denominator by $\sec^2\frac{1}{2}\theta$, this becomes

$$\int\frac{\sec^2\frac{1}{2}\theta\,d\theta}{a+b+(a-b)\tan^2\frac{1}{2}\theta},$$

and, putting for abbreviation

$$\tan\tfrac{1}{2}\theta = y,$$

we have, since $\frac{1}{2}\sec^2\frac{1}{2}\theta\,d\theta = dy$,

$$\int\frac{d\theta}{a+b\cos\theta} = 2\int\frac{dy}{a+b+(a-b)y^2}.$$

The form of this integral depends upon the relative values of a and b. Assuming a to be positive, if b, which may be either positive or negative, is numerically less than a, we may put

$$\frac{a+b}{a-b} = c^2.$$

The integral may then be written in the form

$$\frac{2}{a-b}\int\frac{dy}{c^2+y^2},$$

the value of which is, by formula (k'),

$$\frac{2}{c(a-b)}\tan^{-1}\frac{y}{c}.$$

Hence, substituting their values for y and c, we have, in this case,

$$\int\frac{d\theta}{a+b\cos\theta} = \frac{2}{\sqrt{(a^2-b^2)}}\tan^{-1}\left[\sqrt{\frac{a-b}{a+b}}\tan\tfrac{1}{2}\theta\right]. \quad . \quad (G)$$

If, on the other hand, b is numerically greater than a, this expression for the integral involves imaginary quantities; but putting

$$\frac{b+a}{b-a} = c^2,$$

the integral becomes

$$\frac{2}{b-a}\int\frac{dy}{c^2-y^2},$$

the value of which is, by formula (A'), Art. 17,

$$\frac{1}{c(b-a)}\log\frac{c+y}{c-y}.$$

Therefore, in this case,

$$\int \frac{d\theta}{a + b \cos\theta} = \frac{1}{\sqrt{(b^2 - a^2)}} \log \frac{\sqrt{(b+a)} + \sqrt{(b-a)} \tan \frac{1}{2}\theta}{\sqrt{(b+a)} - \sqrt{(b-a)} \tan \frac{1}{2}\theta}. \quad . \quad (G)$$

36. If $e < 1$, formula (G) of the preceding article gives

$$\int \frac{d\theta}{1 + e \cos\theta} = \frac{2}{\sqrt{(1 - e^2)}} \tan^{-1}\left[\sqrt{\frac{1 - e}{1 + e}} \tan \tfrac{1}{2}\theta\right]. \quad . \quad (1)$$

Putting

$$\sqrt{\frac{1 - e}{1 + e}} \cdot \tan \tfrac{1}{2}\theta = \tan \tfrac{1}{2}\phi, \quad . \quad . \quad . \quad . \quad . \quad (2)$$

and noticing that $\phi = 0$ when $\theta = 0$, we may write

$$\int_0 \frac{d\theta}{1 + e \cos\theta} = \frac{\phi}{\sqrt{(1 - e^2)}}. \quad . \quad . \quad . \quad . \quad . \quad (3)$$

Now, if in equation (1) we put ϕ for θ and change the sign of e, we obtain

$$\int_0 \frac{d\phi}{1 - e \cos\phi} = \frac{2}{\sqrt{(1 - e^2)}} \tan^{-1}\left[\sqrt{\frac{1 + e}{1 - e}} \tan \tfrac{1}{2}\phi\right];$$

hence, by equation (2),

$$\int_0 \frac{d\phi}{1 - e \cos\phi} = \frac{\theta}{\sqrt{(1 - e^2)}}. \quad . \quad . \quad . \quad . \quad . \quad (4)$$

Equations (3) and (4) are equivalent to

$$\frac{d\theta}{1 + e \cos\theta} = \frac{d\phi}{\sqrt{(1 - e^2)}}, \quad . \quad . \quad . \quad . \quad . \quad (5)$$

and

$$\frac{d\phi}{1 - e \cos\phi} = \frac{d\theta}{\sqrt{(1 - e^2)}}, \quad . \quad . \quad . \quad . \quad . \quad (6)$$

the product of which gives

$$(1 + e \cos \theta)\,(1 - e \cos \phi) = 1 - e^2 \quad . \quad . \quad . \quad . \quad (7)$$

By means of these relations any expression of the form

$$\int \frac{d\theta}{(1 + e \cos \theta)^n},$$

where n is a positive integer, may be reduced to an integrable form. For

$$\int \frac{d\theta}{(1 + e \cos \theta)^n} = \int \frac{d\theta}{1 + e \cos \theta}\, \frac{1}{(1 + e \cos \theta)^{n-1}};$$

hence, by equations (5) and (7),

$$\int_0 \frac{d\theta}{(1 + e \cos \theta)^n} = \frac{1}{(1 - e^2)^{n-\frac{1}{2}}} \int_0 (1 - e \cos \phi)^{n-1}\, d\phi .$$

By expanding $(1 - e \cos \phi)^{n-1}$, the last expression is reduced to a series of integrals involving powers of $\cos \phi$; these may be evaluated by the methods given in this section and Section VI, and the results expressed in terms of θ by means of equation (2) or of equation (7).

Examples III.

1. $\int \tan^4 mx\, dx$, $\qquad \dfrac{\tan^3 mx}{3m} - \dfrac{\tan mx}{m} + x.$

2. $\int_0^{\frac{\pi}{4}} \tan^7 x\, dx$, $\qquad \frac{5}{12} - \frac{1}{2} \log 2.$

3. $\int \sec^4 (\theta + \alpha)\, d\theta$, $\qquad \dfrac{\tan^3(\theta + \alpha)}{3} + \tan (\theta + \alpha).$

4. $\int_0^{\frac{\pi}{m}} \sin^3 mx\, dx,$ $\qquad \dfrac{4}{3m}.$

5. $\int \sin^2\theta \cos^3\theta\, d\theta,$ $\qquad \dfrac{\sin^3\theta}{3} - \dfrac{\sin^5\theta}{5}.$

6. $\int \sqrt{(\sin\theta)} \cos^5\theta\, d\theta,$ $\qquad \frac{2}{3}\sin^{\frac{3}{2}}\theta - \frac{4}{7}\sin^{\frac{7}{2}}\theta + \frac{2}{11}\sin^{\frac{11}{2}}\theta.$

7. $\int_0^{\frac{\pi}{2}} \cos^4\theta \sin^3\theta\, d\theta,$ $\qquad \dfrac{2}{35}.$

8. $\int \dfrac{\sin^3\theta\, d\theta}{\sqrt{(\cos\theta)}},$ $\qquad \frac{2}{5}\cos^{\frac{5}{2}}\theta - 2\cos^{\frac{1}{2}}\theta.$

9. $\int \dfrac{d\theta}{\sin^2\theta\cos^2\theta},$ *Multiply by* $\sin^2\theta + \cos^2\theta.$ $\qquad \tan\theta - \cot\theta.$

10. $\int \dfrac{\sin^3 x}{\cos^5 x}\, dx,$ *See Art.* 28. $\qquad \dfrac{\tan^4 x}{4}.$

11. $\int \dfrac{d\theta}{\sin^3\theta\cos^3\theta},$ $\qquad \frac{1}{2}(\tan^2\theta - \cot^2\theta) + 2\log\tan\theta.$

12. $\int \dfrac{\sqrt{(\sin\theta)}\, d\theta}{\cos^{\frac{5}{2}}\theta},$ $\qquad \frac{2}{3}\tan^{\frac{3}{2}}\theta.$

13. $\int \dfrac{\sin^3 x\, dx}{\cos^6 x},$ $\qquad \dfrac{1}{5\cos^5 x} - \dfrac{1}{3\cos^3 x}.$

14. $\int \dfrac{\sin^2 x\, dx}{\cos^6 x},$ $\qquad \dfrac{\tan^5 x}{5} + \dfrac{\tan^3 x}{3}.$

15. $\int \sin^2\theta\cos^2\theta\, d\theta,$ $\qquad \frac{1}{16}[2\theta - \sin 2\theta\cos 2\theta].$

16. $\int_0^{\frac{\pi}{m}} \sin^2 mx\, dx$, $\frac{\pi}{2m}$.

17. $\int \frac{\sin^2 \theta\, d\theta}{\cos\theta}$, $\log\tan\left[\frac{\pi}{4}+\frac{\theta}{2}\right] - \sin\theta$.

18. $\int_{\frac{\pi}{3}}^{\frac{\pi}{2}} \frac{\cos^2\theta\, d\theta}{\sin\theta}$, $\frac{1}{2}(\log 3 - 1)$.

19. $\int \frac{d\theta}{\sin\theta + \cos\theta}$, $\frac{1}{\sqrt{2}}\log\tan\left[\frac{\theta}{2}+\frac{\pi}{8}\right]$.

20. $\int \frac{dx}{1+\cos x}$, $\tan\frac{1}{2}x$.

21. $\int_{\frac{\pi}{2}} \frac{dx}{1-\cos x}$, $1 - \cot\frac{1}{2}x$.

22. $\int \frac{dx}{1 \pm \sin x}$,

Multiply both terms of the fraction by $1 \mp \sin x$. $\tan x \pm \sec x$.

23. $\int \frac{d\theta}{\sec\theta \pm \tan\theta}$, $\log\tan\left[\frac{\pi}{4}+\frac{\theta}{2}\right] \pm \log\cos\theta$.

24. $\int \cos\theta\cos 3\theta\, d\theta$. *See Art.* 33. $\frac{1}{8}\sin 4\theta + \frac{1}{4}\sin 2\theta$.

25. $\int_0^{\frac{\pi}{2}} \cos\theta\cos 2\theta\, d\theta$, $\frac{1}{3}$.

26. $\int_0^{\frac{\pi}{4}} \sin^2\theta\sin 2\theta\, d\theta$, $\frac{1}{2}\sin^4\theta\Big]_0^{\frac{\pi}{4}} = \frac{1}{8}$.

27. $\int_0^{\frac{\pi}{2}} \sin 3\theta \sin 2\theta\, d\theta$, $\frac{2}{5}$.

28. $\int_0 \sin m\theta \cos n\theta \, d\theta,$

$$\frac{1 - \cos(m+n)\theta}{2(m+n)} + \frac{1 - \cos(m-n)\theta}{2(m-n)}.$$

29. $\int \cos x \cos 2x \cos 3x \, dx,$

Reduce products to sums by means of equation (2), *Art.* 33.

$$\frac{1}{4}\left[\frac{\sin 6x}{6} + \frac{\sin 4x}{4} + \frac{\sin 2x}{2} + x\right].$$

30. $\int_0^\pi \sqrt{(1 - \cos x)}\, dx,$ $2\sqrt{2}.$

31. $\int \frac{dx}{a^2 \cos^2 x + b^2 \sin^2 x},$ $\frac{1}{ab}\tan^{-1}\left[\frac{b}{a}\tan x\right].$

32. $\int \frac{dx}{1 + \cos^2 x},$ $\frac{1}{\sqrt{2}}\tan^{-1}\frac{\tan x}{\sqrt{2}}.$

33. $\int \frac{dx}{a^2 \cos^2 x - b^2 \sin^2 x},$ $\frac{1}{2ab}\log\frac{a + b\tan\theta}{a - b\tan\theta}.$

34. $\int \frac{\sin x \, dx}{\sqrt{(3\cos^2 x + 4\sin^2 x)}},$ $\cos^{-1}\{\frac{1}{2}\cos x\}.$

35. $\int \frac{\sin x \cos^3 x \, dx}{1 + a^2 \cos^2 x},$

Putting y *for* $\cos x$, *the integral becomes* $-\int \frac{y^3\, dy}{1 + a^2y^2}.$

$$-\frac{\cos x}{a^2} + \frac{\tan^{-1}(a\cos x)}{a^3}.$$

36. $\displaystyle\int \frac{d\theta}{a + b\sin\theta}.$

Put $\sin\theta = \cos(\theta - \tfrac{1}{2}\pi)$, *and use formulas* (*G*) *and* (*G'*).

$$\text{If } a > b,\ \frac{2}{\sqrt{(a^2 - b^3)}} \tan^{-1}\left[\sqrt{\frac{a - b}{a + b}} \tan\frac{2\theta - \pi}{4}\right].$$

$$\text{If } a < b,\ \frac{1}{\sqrt{(b^2 - a^2)}} \log\frac{\sqrt{(b + a)} + \sqrt{(b - a)}\tan(\tfrac{1}{2}\theta - \tfrac{1}{4}\pi)}{\sqrt{(b + a)} - \sqrt{(b - a)}\tan(\tfrac{1}{2}\theta - \tfrac{1}{4}\pi)}.$$

37. $\displaystyle\int \frac{d\theta}{3 + 5\cos\theta}$, $\qquad \displaystyle\frac{1}{4}\log\frac{2 + \tan\frac{1}{2}\theta}{2 - \tan\frac{1}{2}\theta}.$

38. $\displaystyle\int \frac{d\theta}{5 + 3\cos\theta}$, $\qquad \tfrac{1}{2}\tan^{-1}[\tfrac{1}{2}\tan\tfrac{1}{2}\theta].$

39. $\displaystyle\int \frac{d\theta}{5 - 4\cos\theta}$, $\qquad \tfrac{2}{3}\tan^{-1}\{3\tan\tfrac{1}{2}\theta\}.$

40. $\displaystyle\int \frac{d\theta}{2\cos\theta - 1}$, $\qquad \displaystyle\frac{1}{\sqrt{3}}\log\frac{1 - \sqrt{3}\tan\frac{1}{2}\theta}{1 + \sqrt{3}\tan\frac{1}{2}\theta}.$

41. $\displaystyle\int_0^{\frac{\pi}{2}} \frac{d\theta}{3 - \cos\theta}$, $\qquad \displaystyle\frac{\tan^{-1}\sqrt{2}}{\sqrt{2}}.$

42. $\displaystyle\int_0^{\frac{\pi}{3}} \frac{d\theta}{2 - \cos\theta}$, $\qquad \displaystyle\frac{\pi}{2\sqrt{3}}.$

43. $\displaystyle\int \frac{d\theta}{(1 + e\cos\theta)^2}$, *See Art.* 36.

$$\frac{1}{(1 - e^2)^{\frac{3}{2}}}\cos^{-1}\frac{e + \cos\theta}{1 + e\cos\theta} - \frac{e}{1 - e^2}\,\frac{\sin\theta}{1 + e\cos\theta}.$$

44. $\displaystyle\int_0^{\pi} \frac{d\theta}{(1 + e\cos\theta)^3}$, $\qquad \displaystyle\frac{(2 + e^2)\pi}{2(1 - e^2)^{\frac{5}{2}}}.$

45. $$\int \frac{p\cos x + q\sin x}{a\cos x + b\sin x}\,dx,$$

Solution :—

By adding and subtracting an undetermined constant, the fraction may be written in the form

$$\frac{p\cos x + q\sin x + A\,(a\cos x + b\sin x)}{a\cos x + b\sin x} - A,$$

we may now assume

$$p\cos x + q\sin x + A\,(a\cos x + b\sin x) = k\,(b\cos x - a\sin x);$$

the expression is then readily integrated, and A and k so determined as to make the equation last written an identity. The result is

$$\int \frac{p\cos x + q\sin x}{a\cos x + b\sin x}\,dx = \frac{ap + bq}{a^2 + b^2}\,x + \frac{bp - aq}{a^2 + b^2}\log\,(a\cos x + b\sin x).$$

46. $$\int \frac{dx}{a + b\tan x},$$ See *Ex.* 45.

$$\frac{ax}{a^2 + b^2} + \frac{b}{a^2 + b^2}\log\,(a\cos x + b\sin x).$$

47. Find the area of the ellipse

$$x = a\cos\phi \qquad y = b\sin\phi.$$

$$-\,4ab\int_{\frac{1}{2}\pi}^{0}\sin^2\phi\,d\phi = \pi ab.$$

48. Find the area of the cycloid

$$x = a\,(\psi - \sin\psi) \qquad y = a\,(1 - \cos\psi).$$

$$a^2\int_0^{2\pi}(1 - \cos\psi)^2\,d\psi = 3a^2\pi.$$

49. Find the area of the trochoid $(b < a)$

$$x = a\psi - b\sin\psi \qquad y = a - b\cos\psi.$$

$$(2a^2 + b^2)\,\pi.$$

50. Find the area of the loop, and also the area between the curve and the asymptote, in the case of the strophoid whose polar equation is

$$r = a\,(\sec\theta \pm \tan\theta).$$

Solution :—

Using θ as an auxiliary variable, we have

$$x = a\,(1 \pm \sin\theta) \qquad y = a\left[\tan\theta \pm \frac{\sin^2\theta}{\cos\theta}\right],$$

the upper sign corresponding to the infinite branch, and the lower to the loop. Hence, for the half areas we obtain

$$+\,a^2\int_0^{\frac{1}{2}\pi}\sin\theta\,d\theta + a^2\int_0^{\frac{1}{2}\pi}\sin^2\theta\,d\theta = a^2\left[1 + \frac{\pi}{4}\right]$$

and

$$-a^2\int_{\frac{1}{2}\pi}^{0}\sin\theta\,d\theta + a^2\int_{\frac{1}{2}\pi}^{0}\sin^2\theta\,d\theta = a^2\left[1 - \frac{\pi}{4}\right].$$

CHAPTER II.

METHODS OF INTEGRATION—CONTINUED.

IV.

Integration by Change of Independent Variable.

37. IF x is the independent variable used in expressing an integral, and y is any function of x, the integral may be expressed in terms of y, by substituting for x and dx their values in terms of y and dy. By properly assuming the function y, the integral may frequently be made to take a directly integrable form. For example, the integral

$$\int \frac{x\,dx}{(ax+b)^2}$$

will obviously be simplified by assuming

$$y = ax + b$$

for the new independent variable. This assumption gives

$$x = \frac{y-b}{a}, \qquad \text{whence} \qquad dx = \frac{dy}{a};$$

substituting, we have

$$\int \frac{x\,dx}{(ax+b)^2} = \frac{1}{a^2}\int \frac{(y-b)\,dy}{y^2}$$

$$= \frac{1}{a^2}\log y + \frac{b}{a^2 y};$$

or replacing y by x in the result,

$$\int \frac{x\,dx}{(ax+b)^2} = \frac{1}{a^2}\log(ax+b) + \frac{b}{a^2(ax+b)}.$$

38. Again, if in the integral

$$\int \frac{dx}{\epsilon^x - 1}$$

we put $y = \epsilon^x$, whence

$$x = \log y, \qquad \text{and} \qquad dx = \frac{dy}{y},$$

we have

$$\int \frac{dx}{\epsilon^x - 1} = \int \frac{dy}{y(y-1)}.$$

Hence, by formula (A), Art. 17,

$$\int \frac{dx}{\epsilon^x - 1} = \log \frac{y-1}{y} = \log(\epsilon^x - 1) - x.$$

It is easily seen that, by this change of independent variable, any integral in which the coefficient of dx is a rational function of ϵ^x, may be transformed into one in which the coefficient of dy is a rational function of y.

Transformation of Trigonometric Forms.

39. When in a trigonometric integral the coefficient of $d\theta$ is a rational function of $\tan\theta$, the integral will take a rational algebraic form if we put

$$\tan\theta = x, \qquad \text{whence} \qquad d\theta = \frac{dx}{1+x^2}.$$

For example, by this transformation, we have

$$\int \frac{d\theta}{1 + \tan\theta} = \int \frac{dx}{(1 + x^2)(1 + x)}.$$

Decomposing the fraction in the latter integral, we have

$$\int \frac{d\theta}{1 + \tan\theta} = \frac{1}{2}\int \frac{dx}{1 + x^2} - \frac{1}{2}\int \frac{x\,dx}{1 + x^2} + \frac{1}{2}\int \frac{dx}{1 + x}$$

$$= \tfrac{1}{2}\tan^{-1}x - \tfrac{1}{4}\log(1 + x^2) + \tfrac{1}{2}\log(1 + x)$$

$$= \frac{1}{2}\left[\theta + \log\frac{1 + \tan\theta}{\sec\theta}\right],$$

or $$\int \frac{d\theta}{1 + \tan\theta} = \tfrac{1}{2}[\theta + \log(\cos\theta + \sin\theta)].$$

40. The method given in the preceding article may be employed when the coefficient of $d\theta$ is a *homogeneous rational function of* $\sin\theta$ and $\cos\theta$, *of a degree indicated by an even integer*; for such a function is a rational function of $\tan\theta$. It may also be noticed that, when the coefficient of $d\theta$ is *any rational function* of $\sin\theta$ and $\cos\theta$, the integral becomes rational and algebraic if we put

$$z = \tan\frac{\theta}{2};$$

for this gives

$$\sin\theta = \frac{2z}{1 + z^2}, \qquad \cos\theta = \frac{1 - z^2}{1 + z^2}, \qquad d\theta = \frac{2dz}{1 + z^2}.$$

This transformation has in fact been already employed in the integration of $\dfrac{d\theta}{a + b\cos\theta}$. See Art. 35.

Limits of the Transformed Integral.

41. When a definite integral is transformed by a change of independent variable, it is necessary to make a corresponding change in the limits. If, for example, in the integral

$$\int_a^\infty \frac{dx}{(a^2 + x^2)^2}$$

we put $\quad x = a \tan\theta, \quad$ whence $\quad dx = a \sec^2\theta \, d\theta,$

we must at the same time replace the limits a and ∞, which are values of x, by $\frac{1}{4}\pi$ and $\frac{1}{2}\pi$, the corresponding values of θ. Thus

$$\int_a^\infty \frac{dx}{(a^2 + x^2)^2} = \frac{1}{a^3}\int_{\frac{\pi}{4}}^{\frac{\pi}{2}} \cos^2\theta \, d\theta$$

$$= \frac{1}{2a^3}\Big[\theta + \sin\theta\cos\theta\Big]_{\frac{\pi}{4}}^{\frac{\pi}{2}} = \frac{\pi - 2}{8a^3}.$$

The Reciprocal of x taken as the New Independent Variable.

42. In the case of fractional integrals, it is sometimes useful to take the reciprocal of x as the new independent variable. For example, let the given integral be

$$\int \frac{dx}{x^3 (x + 1)^2}.$$

Putting $\quad x = \frac{1}{y}, \quad$ whence $\quad dx = -\frac{dy}{y^2},$

we have

$$\int \frac{dx}{x^3(x+1)^2} = -\int \frac{y^3\,dy}{y^2\left(1+\frac{1}{y}\right)^2} = -\int \frac{y^3\,dy}{(y+1)^2}.$$

Transforming again by putting $z = y + 1$, the integral becomes

$$-\int \frac{(z-1)^3}{z^2}\,dz = -\int z\,dz + 3\int dz - 3\int \frac{dz}{z} + \int \frac{dz}{z^2}$$

$$= -\frac{z^2}{2} + 3z - 3\log z - \frac{1}{z}.$$

Therefore, since $z = y + 1 = \frac{1}{x} + 1 = \frac{x+1}{x}$,

$$\int \frac{dx}{x^3(x+1)^2} = -\frac{(x+1)^2}{2x^2} + \frac{3(x+1)}{x} - \frac{x}{x+1} - 3\log\frac{x+1}{x}.$$

A Power of x taken as the New Independent Variable.

43. The transformation of an integral by the assumption,

$$y = x^n. \quad . \quad . \quad . \quad . \quad . \quad . \quad . \quad . \quad . \quad . \quad (1)$$

is not generally useful, since the substitution

$$x = y^{\frac{1}{n}}, \qquad \text{whence} \qquad dx = \frac{1}{n}\,y^{\frac{1}{n}-1}\,dy,$$

will usually introduce radicals. Exceptional cases, however, occur. For, since logarithmic differentiation of equation (1) gives

$$\frac{dx}{x} = \frac{dy}{ny}, \quad . \quad . \quad . \quad . \quad . \quad . \quad . \quad . \quad . \quad (2)$$

it is evident that, *if the expression to be integrated is the product of* $\frac{dx}{x}$ *and a function of* x^n, *the transformed expression will be the product of* $\frac{dy}{ny}$ *and the like function of* y.

For example, the expression

$$\frac{(x^4 - 1)\,dx}{x(x^4 + 1)},$$

which is the product of $\frac{dx}{x}$ and a rational function of x^4, becomes

$$\frac{y - 1}{4y(y + 1)}\,dy,$$

a rational function of y. Hence, decomposing the fraction in the latter expression, we have

$$\int \frac{(x^4 - 1)\,dx}{x(x^4 + 1)} = \frac{1}{4}\int \frac{y - 1}{y(y + 1)}\,dy = \frac{1}{4}\log\frac{(y + 1)^2}{y}$$

$$= \log\frac{\sqrt{(x^4 + 1)}}{x}.$$

44. When this method is applied to an integral whose form at the same time suggests the employment of the reciprocal, as in Art. 42, we may at once assume $y = x^{-n}$. Thus, given the integral

$$\int_1^\infty \frac{dx}{x^4(2 + x^8)};$$

putting $y = x^{-3}$, whence $\dfrac{dx}{x} = -\dfrac{dy}{3y}$,

we obtain

$$\int_1^\infty \frac{dx}{x^4(2 + x^3)} = -\frac{1}{3}\int_1^0 \frac{y\,dy}{2y + 1}$$

$$= -\frac{y}{6} + \frac{\log(2y + 1)}{12}\Big]_1^0 = \frac{2 - \log 3}{12}.$$

45. The same mode of transforming may be employed to simplify the coefficient of $\dfrac{dx}{x}$, when this coefficient is not a rational function of x^n. Thus, the integral

$$\int \frac{dx}{x\sqrt{(x^3 - a^3)}}$$

will take the form of the fundamental integral (l'), if we put

$x^3 = y^2$, whence $\dfrac{dx}{x} = \dfrac{2}{3}\dfrac{dy}{y}$.

Making the substitutions, we have

$$\int \frac{dx}{x\sqrt{(x^3 - a^3)}} = \frac{2}{3}\int \frac{dy}{y\sqrt{(y^2 - a^3)}} = \frac{2}{3a^{\frac{3}{2}}}\sec^{-1}\frac{y}{a^{\frac{3}{2}}} = \frac{2}{3a^{\frac{3}{2}}}\sec^{-1}\left(\frac{x}{a}\right)^{\frac{3}{2}}.$$

Examples IV.

2. $\int \frac{x\,dx}{(1-x)^3}$, $\quad \frac{2x-1}{2(1-x)^2}$.

3. $\int \frac{x^2-x+1}{(2x+1)^2}\,dx$, $\quad \frac{2x+1}{8} - \frac{\log(2x+1)}{2} - \frac{7}{8(2x+1)}$.

4. $\int_{-1}^{0} \frac{x^2\,dx}{(x+2)^3}$, $\quad \log y + \frac{4y-2}{y^2}\Big]_1^2 = \log 2 - \frac{1}{2}$.

5. $\int \frac{dx}{1+\varepsilon^x}$ $\quad x - (\log 1 + \varepsilon^x)$.

6. $\int \frac{dx}{\varepsilon^x - \varepsilon^{-x}}$, $\quad \frac{1}{2}\log\frac{\varepsilon^x - 1}{\varepsilon^x + 1}$.

7. $\int_{-\infty}^{0} \frac{\varepsilon^{2x}\,dx}{\varepsilon^x + 1}$, $\quad 1 - \log 2$.

8. $\int \frac{\varepsilon^x + 1}{1 - \varepsilon^{-x}}\,dx$, $\quad \varepsilon^x + 2\log(\varepsilon^x - 1)$.

9. $\int \frac{2+\tan\theta}{3-\tan\theta}\,d\theta$, $\quad \frac{\theta - \log(3\cos\theta - \sin\theta)}{2}$.

10. $\int \frac{d\theta}{\tan^2\theta - 1}$, $\quad \frac{1}{4}\log\frac{\tan\theta - 1}{\tan\theta + 1} - \frac{\theta}{2}$.

11. $\int \frac{\tan^2\theta\,d\theta}{\tan^2\theta - 1}$, $\quad \frac{1}{4}\log\frac{\tan\theta - 1}{\tan\theta + 1} + \frac{\theta}{2}$.

12. $\int \frac{\cos\theta\,d\theta}{a\cos\theta - b\sin\theta}$, $\quad \frac{a\theta - b\log(a\cos\theta - b\sin\theta)}{a^2+b^2}$.

13. $\int \frac{\cos\theta\, d\theta}{\cos(\alpha+\theta)}$, *Put* $\theta' = \alpha + \theta$.

$$(\theta+\alpha)\cos\alpha - \sin\alpha \log\cos(\theta+\alpha).$$

14. $\int \frac{\sin(\theta+\alpha)}{\sin(\theta+\beta)}\, d\theta$,

$$(\theta+\beta)\cos(\alpha-\beta) + \sin(\alpha-\beta)\log\sin(\theta+\beta).$$

15. $\int \tan(\theta+\alpha)\cos\theta\, d\theta$, $\quad -\cos\theta + \sin\alpha \log\tan\frac{2\theta+2\alpha+\pi}{4}$.

16. $\int_0^a \frac{\cos\theta\, d\theta}{\sin(\alpha+\theta)}$, $\quad \cos\alpha\log(2\cos\alpha) + \alpha\sin\alpha$.

17. $\int_0^{\frac{\pi}{3}} \frac{\cos\frac{1}{2}\theta}{\cos\theta}\, d\theta$, $\quad \frac{1}{\sqrt{2}}\log\frac{\sqrt{2}+2\sin\theta'}{\sqrt{2}-2\sin\theta'}\Big]_0^{\frac{\pi}{6}} = \frac{\log(3+2\sqrt{2})}{\sqrt{2}}$.

18. $\int \frac{\sin\frac{1}{2}\theta\, d\theta}{\sin\theta}$, $\quad \log\tan\frac{\pi+\theta}{4}$.

19. $\int \frac{x^3\, dx}{(a^2+x^2)^2}$, $\quad \frac{1}{2}\log(a^2+x^2) + \frac{a^2}{2(a^2+x^2)}$.

20. $\int \frac{dx}{x^3(1+x^2)}$, $\quad \log\frac{\sqrt{(1+x^2)}}{x} - \frac{1}{2x^2}$.

21. $\int_0^\infty \frac{x^2\, dx}{(1+x^2)^3}$, $\quad \frac{1}{4}\int_0^{\frac{\pi}{2}} \sin^2 2\theta\, d\theta = \frac{\pi}{16}$.

22. $\int_1^\infty \frac{dx}{x^3(1+x^2)^2}$, $\quad \frac{3}{4} - \log 2$.

23. $\displaystyle\int \frac{dx}{x^3(x+1)}$, $\qquad -\frac{1}{2x^2} + \frac{1}{x} - \log\frac{x+1}{x}$.

24. $\displaystyle\int \frac{dx}{(1-x)^3 x}$, $\qquad \frac{1}{2(1-x)^2} + \frac{1}{1-x} + \log\frac{x}{1-x}$.

25. $\displaystyle\int_1^\infty \frac{dx}{x^3(x^4+2)}$, $\qquad -\frac{y^2}{4} + \frac{1}{8}\log(2y^2+1)\Big]_1^0 = \frac{2-\log 3}{8}$.

26. $\displaystyle\int \frac{dx}{x(a+bx^4)}$, $\qquad \frac{1}{4a}\log\frac{x^4}{a+bx^4}$.

27. $\displaystyle\int \frac{dx}{x(x^n+a^n)}$, $\qquad \frac{1}{na^n}\log\frac{x^n}{x^n+a^n}$.

28. $\displaystyle\int \frac{(x^3+1)\,dx}{x(x^3-1)}$, $\qquad \frac{2}{3}\log(x^3-1) - \log x$.

29. $\displaystyle\int \frac{dx}{x\sqrt{(x^n-a^n)}}$, $\qquad \frac{2}{na^{\frac{n}{2}}}\sec^{-1}\left(\frac{x}{a}\right)^{\frac{n}{2}}$.

V.

Integrals Containing Radicals.

46. An integral containing a single radical, in which the expression under the radical sign is of the first degree, is *rationalized*, that is, transformed into a rational integral, by taking the radical as the value of the new independent variable. Thus, given the integral

$$\int \frac{dx}{1+\sqrt{(x+1)}},$$

putting $$y = \sqrt{(x+1)},$$

whence $x = y^2 - 1$, and $dx = 2y\,dy$,

we have

$$\int \frac{dx}{1 + \sqrt{(x+1)}} = 2\int \frac{y\,dy}{1+y} = 2\int dy - 2\int \frac{dy}{1+y}$$

$$= 2y - 2\log(1+y)$$

$$= 2\sqrt{(x+1)} - 2\log[1 + \sqrt{(x+1)}].$$

47. The same method evidently applies whenever all the radicals which occur in the integral are powers of a single radical, in which the expression under the radical sign is linear. Thus, in the integral

$$\int_1^2 \frac{dx}{(x-1)^{\frac{2}{3}} + (x-1)^{\frac{1}{2}}},$$

the radicals are powers of $(x-1)^{\frac{1}{6}}$; hence we put $y = (x-1)^{\frac{1}{6}}$, and obtain

$$\int_1^2 \frac{dx}{(x-1)^{\frac{2}{3}} + (x-1)^{\frac{1}{2}}} = 6\int_0^1 \frac{y^5\,dy}{y^4 + y^3}$$

$$= 6\int_0^1 (y-1)\,dy + 6\int_0^1 \frac{dy}{y+1} = -3 + 6\log 2.$$

48. An integral in which a binomial expression occurs under the radical sign can sometimes be reduced to the form considered above by the method of Art. 43. For example, since

$$\int \frac{dx}{x(x^3+1)^{\frac{1}{2}}}$$

fulfils the condition given in Art. 43, when $n = 3$, the quantity under the radical sign may be reduced to the first degree. Hence, in accordance with Art. 46, we may take the radical as the value of the new independent variable. Thus, putting

$$z = (x^3 + 1)^{\frac{1}{4}},$$

whence $\quad x^3 = z^4 - 1, \quad$ and $\quad \dfrac{dx}{x} = \dfrac{4z^3\,dz}{3(z^4 - 1)},$

we have

$$\int \frac{dx}{x(x^3 + 1)^{\frac{1}{4}}} = \frac{4}{3}\int \frac{z^2\,dz}{z^4 - 1}.$$

Decomposing the fraction in the latter integral as in Art. 20, we have finally

$$\int \frac{dx}{x(x^3 + 1)^{\frac{1}{4}}} = \frac{2}{3}\tan^{-1}\left[(x^3 + 1)^{\frac{1}{4}}\right] + \frac{1}{3}\log\frac{(x^3 + 1)^{\frac{1}{4}} - 1}{(x^3 + 1)^{\frac{1}{4}} + 1}.$$

Radicals of the Form $\sqrt{(ax^2 + b)}$.

49. It is evident that the method given in the preceding article is applicable to all integrals of the general form

$$\int x^{2m+1}(ax^2 + b)^{n+\frac{1}{2}}\,dx, \quad \ldots \ldots \quad (1)$$

in which m and n are positive or negative integers. These integrals are therefore rationalized by putting

$$y = \sqrt{(ax^2 + b)}.$$

Putting $m = 0$, the form (1) includes the directly integrable case

$$\int (ax^2 + b)^{n+\frac{1}{2}} x\,dx.$$

50. As an illustration let us take the integral

$$\int \frac{dx}{x\sqrt{(x^2 + a^2)}};$$

putting $$y = \sqrt{(x^2 + a^2)},$$

whence $$x^2 = y^2 - a^2, \qquad \text{and} \qquad \frac{dx}{x} = \frac{y\,dy}{y^2 - a^2},$$

we have

$$\int \frac{dx}{x\sqrt{(x^2 + a^2)}} = \int \frac{dy}{y^2 - a^2}.$$

Hence, by equation (A') Art. 17,

$$\int \frac{dx}{x\sqrt{(x^2 + a^2)}} = \frac{1}{2a} \log \frac{y - a}{y + a} = \frac{1}{2a} \log \frac{\sqrt{(x^2 + a^2)} - a}{\sqrt{(x^2 + a^2)} + a}.$$

Rationalizing the denominator of the fraction in this result, we have

$$\frac{\sqrt{(x^2 + a^2)} - a}{\sqrt{(x^2 + a^2)} + a} = \frac{[\sqrt{(x^2 + a^2)} - a]^2}{x^2}.$$

Therefore

$$\int \frac{dx}{x\sqrt{(x^2 + a^2)}} = \frac{1}{a} \log \frac{\sqrt{(x^2 + a^2)} - a}{x}. \quad . \quad . \quad . \quad (H)$$

In a similar manner we may prove that

$$\int \frac{dx}{x\sqrt{(a^2 - x^2)}} = \frac{1}{a} \log \frac{a - \sqrt{(a^2 - x^2)}}{x} \quad . \quad . \quad . \quad . \quad (I)$$

51. Integrals of the form

$$\int x^{2m}(ax^2 + b)^{n + \frac{1}{2}}\, dx \quad . \quad . \quad . \quad . \quad . \quad . \quad . \quad (2)$$

are reducible to the form (1) Art. 49, by first putting $y = \frac{1}{x}$. For example:

$$\int \frac{dx}{(ax^2 + b)^{\frac{3}{2}}}$$

is of the form (2); but, putting $x = \frac{1}{y}$, whence

$$\sqrt{(ax^2 + b)} = \frac{\sqrt{(a + by^2)}}{y} \qquad \text{and} \qquad dx = -\frac{dy}{y^2},$$

we obtain

$$\int \frac{dx}{(ax^2 + b)^{\frac{3}{2}}} = -\int \frac{y\, dy}{(a + by^2)^{\frac{3}{2}}}.$$

The resulting expression is in this case directly integrable. Thus

$$\int \frac{dx}{(ax^2 + b)^{\frac{3}{2}}} = \frac{1}{b\sqrt{(a + by^2)}} = \frac{x}{b\sqrt{(ax^2 + b)}} \quad . \quad . \quad . \quad (J)$$

Integration of $\frac{dx}{\sqrt{(x^2 \pm a^2)}}$.

52. If we assume a new variable z connected with x by the relation

$$z - x = \sqrt{(x^2 \pm a^2)}, \quad . \quad . \quad . \quad . \quad . \quad . \quad . \qquad (1)$$

we have, by squaring,

$$z^2 - 2zx = \pm a^2, \quad . \quad . \quad . \quad . \quad . \quad . \quad . \quad . \qquad (2)$$

and, by differentiating this equation,

$$2(z - x)\,dz - 2z\,dx = 0;$$

whence

$$\frac{dx}{z - x} = \frac{dz}{z},$$

or by equation (1),

$$\frac{dx}{\sqrt{(x^2 \pm a^2)}} = \frac{dz}{z}. \quad . \quad . \quad . \quad . \quad . \quad . \quad . \qquad (3)$$

Integrating equation (3), we obtain

$$\int \frac{dx}{\sqrt{(x^2 \pm a^2)}} = \log z = \log\left[x + \sqrt{(x^2 \pm a^2)}\right]. \quad . \quad . \quad . \qquad (K)$$

53. Since the value of x in terms of z, derived from equation (2) of the preceding article, is rational, it is obvious that this transformation may be employed to rationalize any expression which consists of the product of $\frac{dx}{\sqrt{(x^2 \pm a^2)}}$ and a rational function of x. For example, let us find the value of

$$\int \sqrt{(x^2 \pm a^2)}\,dx,$$

which may be written in the form

$$\int (x^2 \pm a^2) \frac{dx}{\sqrt{(x^2 \pm a^2)}}.$$

By equation (2)

$$x = \frac{z^2 \mp a^2}{2z}, \quad . \quad . \quad . \quad . \quad . \quad . \quad . \quad . \quad (4)$$

whence

$$x^2 \pm a^2 = \frac{(z^2 \pm a^2)^2}{4z^2}. \quad . \quad . \quad . \quad . \quad . \quad . \quad . \quad (5)$$

Therefore, by equations (3) and (5),

$$\int \sqrt{(x^2 \pm a^2)}\, dx = \frac{1}{4} \int \frac{(z^2 \pm a^2)^2}{z^3}\, dz$$

$$= \frac{1}{4} \int z\, dz \pm \frac{a^2}{2} \int \frac{dz}{z} + \frac{a^4}{4} \int \frac{dz}{z^3}$$

$$= \frac{z^4 - a^4}{8z^2} \pm \frac{a^2}{2} \log z.$$

By equations (4) and (5), the first term of the last member is equal to $\frac{1}{2} x \sqrt{(x^2 \pm a^2)}$. Hence

$$\int \sqrt{(x^2 \pm a^2)}\, dx = \frac{x\sqrt{(x^2 \pm a^2)}}{2} \pm \frac{a^2}{2} \log [x + \sqrt{(x^2 \pm a^2)}]. \quad . \quad (L)$$

Transformation to Trigonometric Forms.

54. Integrals involving either of the radicals

$$\sqrt{(a^2 - x^2)}, \qquad \sqrt{(a^2 + x^2)}, \qquad \text{or} \qquad \sqrt{(x^2 - a^2)}$$

can be transformed into rational trigonometric integrals. The transformation is effected in the first case by putting

$$x = a \sin \theta, \qquad \text{whence} \qquad \sqrt{(a^2 - x^2)} = a \cos \theta;$$

in the second case, by putting

$$x = a \tan \theta, \qquad \text{whence} \qquad \sqrt{(a^2 + x^2)} = a \sec \theta;$$

and in the third case, by putting

$$x = a \sec \theta, \qquad \text{whence} \qquad \sqrt{(x^2 - a^2)} = a \tan \theta.$$

55. As an illustration, let us take the integral

$$\int \sqrt{(a^2 - x^2)}\, dx;$$

putting $x = a \sin \theta$, we have $\sqrt{(a^2 - x^2)} = a \cos \theta$, $dx = a \cos \theta\, d\theta$; hence

$$\int \sqrt{(a^2 - x^2)}\, dx = a^2 \int \cos^2 \theta\, d\theta$$

$$= \frac{a^2 \theta}{2} + \frac{a^2 \sin \theta \cos \theta}{2},$$

by formula (C) Art. 29. Replacing θ by x in the result,

$$\int \sqrt{(a^2 - x^2)}\, dx = \frac{a^2}{2} \sin^{-1} \frac{x}{a} + \frac{x \sqrt{(a^2 - x^2)}}{2}. \quad . \quad . \quad . \quad (M)$$

Regarding the radical as a positive quantity, the value of θ may be restricted to the *primary* value of the symbol $\sin^{-1} \frac{x}{a}$ (see Diff. Calc., Art. 54); that is, as x passes from $-a$ to $+a$, θ passes from $-\frac{1}{2}\pi$ to $+\frac{1}{2}\pi$.

Radicals of the Form $\sqrt{(ax^2 + bx + c)}$.

56. When a radical of the form $\sqrt{(ax^2 + bx + c)}$ occurs in an integral, a simple change of independent variable will cause the radical to assume one of the forms considered in the preceding articles. Thus, if the coefficient of x^2 is positive,

$$\sqrt{(ax^2 + bx + c)} = \sqrt{a}\sqrt{\left[\left(x + \frac{b}{2a}\right)^2 + \frac{4ac - b^2}{4a^2}\right]},$$

in which, if we put $x + \dfrac{b}{2a} = y$, the radical takes the form $\sqrt{(y^2 + a^2)}$ or $\sqrt{(y^2 - a^2)}$, according as $4ac - b^2$ is positive or negative. If a is negative, the radical can in like manner be reduced to the form $\sqrt{(a^2 - y^2)}$ or $\sqrt{(-a^2 - y^2)}$; but the latter will never occur, since it is imaginary for all values of y, and therefore imaginary for all values of x.

For example, by this transformation, the integral

$$\int \frac{dx}{(ax^2 + bx + c)^{\frac{3}{2}}}$$

can be reduced at once to the form ($\mathcal{J}$), Art. 51. Thus

$$\int \frac{dx}{(ax^2 + bx + c)^{\frac{3}{2}}} = \int \frac{dx}{\left[a\left(x + \frac{b}{2a}\right)^2 + \frac{4ac - b^2}{4a}\right]^{\frac{3}{2}}}$$

$$= \frac{x + \dfrac{b}{2a}}{\dfrac{4ac - b^2}{4a}\sqrt{(ax^2 + bx + c)}} = \frac{4ax + 2b}{(4ac - b^2)\sqrt{(ax^2 + bx + c)}}.$$

57. When the form of the integral suggests a further change of independent variable, we may at once assume the expression for the new variable in the required form. For example, given the integral

$$\int \sqrt{(2ax - x^2)}\, x\, dx;$$

we have $\sqrt{(2ax - x^2)} = \sqrt{[a^2 - (x - a)^2]}$

hence (see Art. 54), if we put $x - a = a \sin\theta$, we have

$$\sqrt{(2ax - x^2)} = a\cos\theta,$$

$$x = a(1 + \sin\theta), \qquad dx = a\cos\theta\, d\theta;$$

$$\therefore \int \sqrt{(2ax - x^2)}\, x\, dx = a^3 \int \cos^2\theta(1 + \sin\theta)\, d\theta$$

$$= \frac{a^3}{2}(\theta + \sin\theta\cos\theta) - \frac{a^3}{3}\cos^3\theta$$

$$= \frac{a^3}{2}\sin^{-1}\frac{x-a}{a} + \frac{a}{2}(x-a)\sqrt{(2ax - x^2)} - \frac{1}{3}(2ax - x^2)^{\frac{3}{2}}$$

$$= \frac{a^3}{2}\sin^{-1}\frac{x-a}{a} + \frac{1}{6}\sqrt{(2ax - x^2)}\,[2x^2 - ax - 3a^2].$$

The Integrals

$$\int\frac{dx}{\sqrt{[(x-\alpha)(x-\beta)]}} \quad \textit{and} \quad \int\frac{dx}{\sqrt{[(x-\alpha)(\beta-x)]}}.$$

58. An integral of the form $\int\frac{dx}{\sqrt{(ax^2 + bx + c)}}$ may by the method of Art. 56, be reduced to the form (K), Art. 52, or to the form (j'), Art. 10, according as a is positive or negative.

But when the quantity under the radical sign can be resolved into linear factors, the formulas deduced below give the value of the integral in forms which are sometimes more convenient.

If α and β are the roots of the equation

$$ax^2 + bx + c = 0,$$

the integral may be put in the form

$$\frac{1}{\sqrt{a}}\int\frac{dx}{\sqrt{[(x-\alpha)(x-\beta)]}} \quad \text{or} \quad \frac{1}{\sqrt{(-a)}}\int\frac{dx}{\sqrt{[(x-\alpha)(\beta-x)]}},$$

according as a is positive or negative. Assuming

$$\sqrt{(x-\alpha)} = z, \qquad \text{whence} \qquad x = z^2 + \alpha \qquad \text{and} \qquad dx = 2z\,dz,$$

we have

$$\int\frac{dx}{\sqrt{[(x-\alpha)(x-\beta)]}} = 2\int\frac{dz}{\sqrt{(z^2+\alpha-\beta)}} = 2\log[z+\sqrt{(z^2+\alpha-\beta)}],$$

by formula (K), Art. 52; hence

$$\int\frac{dx}{\sqrt{[(x-\alpha)(x-\beta)]}} = 2\log[\sqrt{(x-\alpha)}+\sqrt{(x-\beta)}] \quad . \quad . \quad . \quad (N)$$

In like manner we have

$$\int\frac{dx}{\sqrt{[(x-\alpha)(\beta-x)]}} = 2\int\frac{dz}{\sqrt{(\beta-\alpha-z^2)}} = 2\sin^{-1}\frac{z}{\sqrt{(\beta-\alpha)}},$$

by formula (j'); hence

$$\int\frac{dx}{\sqrt{[(x-\alpha)(\beta-x)]}} = 2\sin^{-1}\sqrt{\frac{x-\alpha}{\beta-\alpha}}. \quad . \quad . \quad . \quad (O)$$

It can be shown that the values given in formulas (N) and (O) differ only by constants from the results derived by employing the process given in Art. 56.

Examples V.

1. $\int \sqrt{(a-x)} \cdot x \, dx$, $\qquad -\frac{2}{15}(a-x)^{\frac{3}{2}}(3x+2a)$.

2. $\int \sqrt{(x+a)} \cdot x^2 \, dx$, $\qquad \frac{2}{7}(a+x)^{\frac{7}{2}} - \frac{4a}{5}(a+x)^{\frac{5}{2}} + \frac{2a^2}{3}(a+x)^{\frac{3}{2}}$.

3. $\int \frac{x\,dx}{1+\sqrt{x}}$, $\qquad \frac{2}{3}x^{\frac{3}{2}} - x + 2\sqrt{x} - 2\log(1+\sqrt{x})$.

4. $\int \frac{x\,dx}{\sqrt{(x+a)}}$, $\qquad \frac{2}{3}(x-2a)\sqrt{(x+a)}$.

5. $\int \frac{dx}{\sqrt{x}-1}$, $\qquad 2\sqrt{x} + 2\log(1-\sqrt{x})$.

6. $\int_{-a}^{0}(a+x)^{\frac{1}{3}} x \, dx$, $\qquad \left.\frac{3y^7}{7} - \frac{3ay^4}{4}\right]_0^{a^{\frac{1}{3}}} = -\frac{9a^{\frac{7}{3}}}{28}$.

7. $\int \frac{dx}{x\sqrt{(2ax-a^2)}}$, $\qquad \frac{2}{a}\tan^{-1}\sqrt{\frac{2x-a}{a}}$.

8. $\int_0^a (a-x)^{\frac{3}{2}} x^2 \, dx$, $\qquad \left.-\frac{2y^9}{9} + \frac{4ay^7}{7} - \frac{2a^2y^5}{5}\right]_{\sqrt{a}}^{0} = \frac{16a^{\frac{9}{2}}}{315}$.

9. $\int \frac{dx}{2x^{\frac{1}{2}} - x^{\frac{1}{3}}}$, $\qquad x^{\frac{1}{2}} + \frac{3x^{\frac{1}{3}}}{4} + \frac{3x^{\frac{1}{6}}}{4} + \frac{3\log(2x^{\frac{1}{6}}-1)}{8}$.

10. $\int_0^1 (x+1)^{\frac{2}{3}}\, x\, dx$, $\qquad \left.\frac{3y^8}{8} - \frac{3y^5}{5}\right]_1^{\sqrt[3]{2}} = \frac{3\sqrt[3]{4}}{10} + \frac{9}{40}$.

11. $\int \frac{1-3x}{\sqrt{(1-x)}}\, dx$, $\qquad 2(1+x)\sqrt{(1-x)}$.

12. $\int \frac{x\, dx}{x - \sqrt{(x^2-a^2)}}$, *Rationalize the denominator.*

$$\frac{x^3 + (x^2-a^2)^{\frac{3}{2}}}{3a^2}.$$

13. $\int \frac{dx}{\sqrt{(x+a)} + \sqrt{(x+b)}}$, $\qquad \frac{2(x+a)^{\frac{3}{2}} - 2(x+b)^{\frac{3}{2}}}{3(a-b)}$.

14. $\int \frac{dx}{x\sqrt{(x^4+1)}}$, $\qquad \frac{1}{4}\log\frac{\sqrt{(x^4+1)}-1}{\sqrt{(x^4+1)}+1}$.

15. $\int \frac{\sqrt{(x^4+1)}\, dx}{x}$, $\qquad \frac{\sqrt{(x^4+1)}}{2} + \frac{1}{4}\log\frac{\sqrt{(x^4+1)}-1}{\sqrt{(x^4+1)}+1}$.

16. $\int \frac{(x^n+1)(x^n-1)^{\frac{3}{2}}}{x}\, dx$,

$$\frac{2}{n}\left[\frac{(x^n-1)^{\frac{5}{2}}}{5} + \frac{(x^n-1)^{\frac{3}{2}}}{3} - (x^n-1)^{\frac{1}{2}} + \tan^{-1}\sqrt{(x^n-1)}\right].$$

17. $\int_0^a \frac{x^5\, dx}{(x^2+a^2)^{\frac{3}{2}}}$, $\qquad \left.\frac{y^3}{3} - 2a^2 y - \frac{a^4}{y}\right]_a^{a\sqrt{2}} = \left[\frac{8}{3} - \frac{11\sqrt{2}}{6}\right] a^3$.

18. $\int \frac{\sqrt{(x^2-a^2)}}{x}\, dx \left[= \int \frac{x^2-a^2}{x\sqrt{(x^2-a^2)}}\, dx\right]$,

$$\sqrt{(x^2-a^2)} - a\sec^{-1}\frac{x}{a}.$$

19. $\displaystyle\int \frac{x^2\,dx}{\sqrt{(x^2 + a^2)}}$,

$$\left[= \int \frac{x^2 + a^2 - a^2}{\sqrt{(x^2 + a^2)}}\,dx. \quad \textit{See formulas (L) and (K).}\right]$$

$$\frac{1}{2}x\sqrt{(x^2 + a^2)} - \frac{1}{2}a^2 \log\left[x + \sqrt{(x^2 + a^2)}\right].$$

20. $\displaystyle\int \frac{\sqrt{(a^2 - x^2)}}{x}\,dx, \qquad a \log \frac{a - \sqrt{(a^2 - x^2)}}{x} + \sqrt{(a^2 - x^2)}.$

21. $\displaystyle\int \frac{dx}{x + \sqrt{(x^2 + a^2)}}$,

$$\frac{x}{2a^2}\sqrt{(x^2 + a^2)} + \frac{1}{2}\log\left[x + \sqrt{(x^2 + a^2)}\right] - \frac{x^2}{2a^2}.$$

22. $\displaystyle\int \frac{\sqrt{(x^2 + a^2)}}{x^2}\,dx$, *See Art.* 51.

$$\log\left[\sqrt{(x^2 + a^2)} + x\right] - \frac{\sqrt{(x^2 + a^2)}}{x}.$$

23. $\displaystyle\int \frac{dx}{\sqrt{(x^2 + a^2)} - a}$,

$$\log\left[\sqrt{(x^2 + a^2)} + x\right] - \frac{\sqrt{(x^2 + a^2)}}{x} - \frac{a}{x}.$$

24. $\displaystyle\int \frac{x\,dx}{\sqrt{(1 + x^4)}}$. *See Formula (K).* $\displaystyle\frac{1}{2}\log\left[x^2 + \sqrt{(1 + x^4)}\right].$

25. $\int \sqrt{(ax^2 + b)}\,dx,$ $[a > 0]$ *Put* $\sqrt{(ax^2 + b)} = z - x\sqrt{a}.$

$$\frac{b}{2\sqrt{a}} \log\left[x\sqrt{a} + \sqrt{(ax^2 + b)}\right] + \frac{1}{2}\, x\sqrt{(ax^2 + b)}.$$

26. $\int \frac{dx}{(a + x)\sqrt{(x^2 + b^2)}},$

$$\frac{1}{\sqrt{(a^2 + b^2)}} \log \frac{x + \sqrt{(x^2 + b^2)} + a - \sqrt{(a^2 + b^2)}}{x + \sqrt{(x^2 + b^2)} + a + \sqrt{(a^2 + b^2)}}.$$

27. $\int \frac{dx}{x^2\sqrt{(1 + x^2)}},$ $-\frac{\sqrt{(1 + x^2)}}{x}.$

28. $\int_{\frac{1}{2}}^{1} \frac{dx}{x^2\sqrt{(1 - x^2)}},$ $-\cot\theta\Big]_{\frac{\pi}{6}}^{\frac{\pi}{2}} = \sqrt{3}.$

29. $\int \frac{x^3\,dx}{(x^2 - a^2)^{\frac{3}{2}}},$ $\frac{x^2 - 2a^2}{\sqrt{(x^2 - a^2)}}.$

30. $\int \frac{dx}{(p + qx)\sqrt{(x^2 + 1)}},$

$$\frac{1}{\sqrt{(p^2 + q^2)}} \log\tan\frac{1}{2}\left[\tan^{-1} x + \tan^{-1}\frac{p}{q}\right].$$

31. $\int \frac{dx}{x^3\sqrt{(x^2 - 1)}},$ $-\frac{\sqrt{(x^2 - 1)}}{2x^2} + \frac{1}{2}\sec^{-1} x.$

32. $\int_0^a \frac{x^2\,dx}{(x^2 + a^2)^{\frac{3}{2}}}$ $\log\tan\frac{3\pi}{8} - \frac{\sqrt{2}}{2}.$

33. $\int \frac{dx}{x^4 \sqrt{(x^2-1)}}$, $\qquad (2x^2+1)\frac{\sqrt{(x^2-1)}}{3x^3}$.

34. $\int \frac{dx}{(a-x)\sqrt{(a^2-x^2)}}$, $\qquad \frac{1}{a}\sqrt{\frac{a+x}{a-x}}$.

35. $\int \frac{dx}{(1+x^2)\sqrt{(1-x^2)}}$, $\qquad \frac{1}{\sqrt{2}}\tan^{-1}\frac{x\sqrt{2}}{\sqrt{(1-x^2)}}$.

36. $\int_0^a \frac{x^{\frac{5}{2}}\,dx}{\sqrt{(a-x)}} \left[= \int_0^a \frac{x^3\,dx}{\sqrt{(ax-x^2)}}\right]$

$$\left(\frac{1}{2}a\right)^3 \int_{-\frac{\pi}{2}}^{\frac{\pi}{2}} (1+\sin\theta)^3\,d\theta = \frac{5\pi a^3}{16}$$

37. $\int \frac{dx}{x\sqrt{(2ax-x^2)}}$, $\qquad -\frac{1}{a}\sqrt{\frac{2a-x}{x}}$.

38. $\int \frac{dx}{x^3\sqrt{(x^4-1)}}$. *Put* $x^2 = z = \sec\theta$. $\qquad \frac{\sqrt{(x^4-1)}}{2x^2}$.

39. $\int_0^a \sqrt{(2ax-x^2)}\cdot dx$, $\qquad \frac{a^2\pi}{4}$.

40. $\int_0^a \sqrt{(2ax-x^2)}\cdot x\,dx$,

$$a^3 \int_{-\frac{\pi}{2}}^{0} \cos^2\theta\,(1+\sin\theta)\,d\theta = a^3\left[\frac{\pi}{4}-\frac{1}{3}\right].$$

41. $\int_0^a \sqrt{(2ax-x^2)}\cdot x^2\,dx$,

$$a^4 \int_{-\frac{\pi}{2}}^{0} \cos^2\theta\,(1+\sin\theta)^2\,d\theta = a^4\left[\frac{5\pi}{16}-\frac{2}{3}\right].$$

42. $\int \frac{dx}{\sqrt{(2ax + x^2)}}$,

by Art. 56, $\log[x + a + \sqrt{(2ax + x^2)}] + C$;

by Art. 58, $\log[\sqrt{x} + \sqrt{(2a + x)}] + C'$.

43. $\int \frac{x\,dx}{\sqrt{(2ax + x^2)}}$, $\sqrt{(2ax + x^2)} - a \log[x + a + \sqrt{(2ax + x^2)}]$.

44. $\int \sqrt{\frac{x}{2a - x}}\,dx \left[= \int \frac{x\,dx}{\sqrt{(2ax - x^2)}} \right]$,

$a \sin^{-1} \frac{x - a}{a} - \sqrt{(2ax - x^2)}$.

45. $\int \frac{dx}{\sqrt{(5 + 4x - x^2)}}$, *by Art.* 56, $\sin^{-1} \frac{x - 2}{3} + C$;

by Art. 58, $2 \sin^{-1} \sqrt{\frac{x + 1}{6}} + C'$.

46. $\int_0^a \frac{dx}{\sqrt{(ax - x^2)}}$, $2 \sin^{-1} \sqrt{\frac{x}{a}} \Big]_0^a = \pi$.

47. $\int \frac{x^2\,dx}{\sqrt{(3 + 2x - x^2)}}$, $3 \sin^{-1} \frac{x - 1}{2} - \frac{(x + 3)\sqrt{(3 + 2x - x^2)}}{2}$.

48. $\int_{-2}^{-\frac{1}{2}} \frac{dx}{\sqrt{(2 - x - x^2)}}$, $\frac{\pi}{2}$.

49. $\int_a^{2a} \frac{dx}{\sqrt{(x^2 - ax)}}$, $\log(3 + 2\sqrt{2})$.

50. Find the area included by the rectangular hyperbola

$$y^2 = 2ax + x^2,$$

and the double ordinate of the point for which $x = 2a$.

$$a^2[6\sqrt{2} - \log(3 + 2\sqrt{2})].$$

51. Find the area included between the cissoid

$$x(x^2 + y^2) = 2ay^2$$

and the coördinates of the point (a, a); also the whole area between the curve and its asymptote.

$$\left(\frac{3}{4}\pi - 2\right)a^2, \qquad \text{and} \qquad 3\pi a^2.$$

52. Find the area of the loop of the strophoid

$$x(x^2 + y^2) + a(x^2 - y^2) = 0;$$

also the area between the curve and its asymptote.

$$2a^2\left(1 - \frac{\pi}{4}\right), \qquad \text{and} \qquad 2a^2\left(1 + \frac{\pi}{4}\right).$$

For the loop put $y = -x\dfrac{a + x}{\sqrt{(a^2 - x^2)}}$, *since* x *is negative between the limits* $-a$ *and* 0.

53. Show that the area of the segment of an ellipse between the minor axis and any double ordinate is $ab\sin^{-1}\dfrac{x}{a} + xy$.

VI.

Integration by Parts.

59. Let u and v be any two functions of x; then since

$$d(uv) = u\,dv + v\,du,$$

$$uv = \int u\,dv + \int v\,du,$$

whence

$$\int u\,dv = uv - \int v\,du \quad . \quad . \quad . \quad . \quad . \quad . \quad . \quad (1)$$

By means of this formula, the integration of an expression of the form $u\,dv$, in which dv is the differential of a known function v, may be made to depend upon the integration of the expression $v\,du$. For example, if

$$u = \cos^{-1} x \qquad \text{and} \qquad dv = dx,$$

we have

$$du = -\frac{dx}{\sqrt{(1 - x^2)}};$$

hence, by equation (1),

$$\int \cos^{-1} x \cdot dx = x \cos^{-1} x + \int \frac{x\,dx}{\sqrt{(1 - x^2)}},$$

in which the new integral is directly integrable; therefore

$$\int \cos^{-1} x \cdot dx = x \cos^{-1} x - \sqrt{(1 - x^2)}.$$

The employment of this formula is called *integration by parts.*

Geometrical Illustration.

60. The formula for integration by parts may be geometrically illustrated as follows. Assuming rectangular axes, let the curve be constructed in which the abscissa and ordinate of each point are corresponding values of v and u, and let this curve cut one of the axes in B. From any point P of this curve draw PR and PS, perpendicular to the axes. Now the area $PBOR$ is a value of the indefinite integral $\int u\,dv$, and in like manner the area PBS is a value of $\int v\,du$;

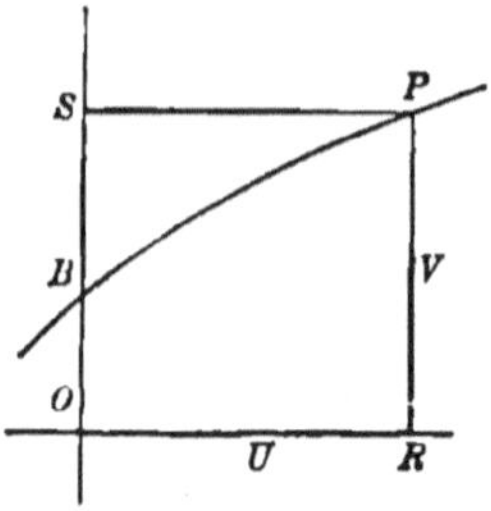

FIG. 2.

and we have

$$\text{Area } PBOR = \text{Rectangle } PSOR - \text{Area } PBS;$$

therefore

$$\int u\,dv = uv - \int v\,du.$$

Applications.

61. In general there will be more than one possible method of selecting the factors u and dv. The latter of course includes the factor dx, but it will generally be advisable to include in it any other factors which permit the direct integration of dv. After selecting the factors, it will be found convenient at once to write the product $u \cdot v$, separating the factors by a period; this will serve as a guide in forming the product

$v\,du$, which is to be written under the integral sign. Thus, let the given integral be

$$\int x^2 \log x \, dx.$$

Taking $x^2 dx$ as the value of dv, since we can integrate this expression directly, we have

$$\int x^2 \log x \, dx = \log x \cdot \frac{1}{3} x^3 - \frac{1}{3}\int x^3 \frac{dx}{x}$$

$$= \frac{1}{3} x^3 \log x - \frac{1}{3}\int x^2 \, dx$$

$$= \frac{x^3}{9}(3 \log x - 1).$$

62. The form of the new integral may be such that a second application of the formula is required before a directly integrable form is produced. For example, let the given integral be

$$\int x^2 \cos x \, dx.$$

In this case we take $\cos x \, dx = dv$; so that having $x^2 = u$, the new integral will contain a lower power of x: thus

$$\int x^2 \cos x \, dx = x^2 \cdot \sin x - 2\int x \sin x \, dx.$$

Making a second application of the formula, we have

$$\int x^2 \cos x \, dx = x^2 \sin x - 2\left[x(-\cos x) + \int \cos x \, dx\right]$$

$$= x^2 \sin x + 2x \cos x - 2 \sin x.$$

63. The method of integration by parts is sometimes employed with advantage, even when the new integral is no simpler than the given one; for, in the process of successive applications of the formula, the original integral may be reproduced, as in the following example:

$$\int \varepsilon^{mx} \sin(nx + \alpha)\, dx$$

$$= \varepsilon^{mx} \cdot \frac{-\cos(nx + \alpha)}{n} + \frac{m}{n}\int \varepsilon^{mx} \cos(nx + \alpha)\, dx$$

$$= -\frac{\varepsilon^{mx}\cos(nx+\alpha)}{n} + \frac{m}{n}\varepsilon^{mx}\frac{\sin(nx+\alpha)}{n} - \frac{m^2}{n^2}\int \varepsilon^{mx}\sin(nx+\alpha)\,dx,$$

in which the integral in the second member is identical with the given integral; hence, transposing and dividing,

$$\int \varepsilon^{mx} \sin(nx + \alpha)\, dx = \frac{\varepsilon^{mx}}{m^2 + n^2}[m\sin(nx + \alpha) - n\cos(nx + \alpha)].$$

64. In some cases it is necessary to employ some other mode of transformation, in connection with the method of parts. For example, given the integral

$$\int \sec^3 \theta\, d\theta;$$

taking $dv = \sec^2 \theta\, d\theta$, we have

$$\int \sec^3 \theta\, d\theta = \sec\theta \cdot \tan\theta - \int \sec\theta \tan^2\theta\, d\theta. \quad . \quad . \quad (1)$$

If now we apply the method of parts to the new integral, by putting

$$\sec\theta\tan\theta\,d\theta = dv,$$

the original integral will indeed be reproduced in the second member; but it will disappear from the equation, the result being an identity. If, however, in equation (1), we transform the final integral by means of the equation $\tan^2\theta = \sec^2\theta - 1$, we have

$$\int\sec^3\theta\,d\theta = \sec\theta\tan\theta - \int\sec^3\theta\,d\theta + \int\sec\theta\,d\theta.$$

Transposing,

$$2\int\sec^3\theta\,d\theta = \frac{\sin\theta}{\cos^2\theta} + \int\frac{d\theta}{\cos\theta};$$

hence, by formula (F), Art. 31,

$$\int\sec^3\theta\,d\theta = \frac{\sin\theta}{2\cos^2\theta} + \frac{1}{2}\log\tan\left[\frac{\pi}{4} + \frac{\theta}{2}\right].$$

Formulas of Reduction.

65. It frequently happens that the new integral introduced by applying the method of parts differs from the given integral only in the values of certain constants. If these constants are expressed algebraically, the formula expressing the first transformation is adapted to the successive transformations of the new integrals introduced, and is called *a formula of reduction.*

For example, applying the method of parts to the integral

$$\int x^n \epsilon^{ax}\, dx,$$

we have

$$\int x^n \epsilon^{ax}\, dx = x^n \cdot \frac{\epsilon^{ax}}{a} - \frac{n}{a}\int x^{n-1} \epsilon^{ax}\, dx, \quad . \quad . \quad . \quad . \quad (1)$$

in which the new integral is of the same form as the given one, the exponent of x being decreased by unity. Equation (1) is therefore a formula of reduction for this function. Supposing n to be a positive integer, we shall finally arrive at the integral $\int \epsilon^{ax}\, dx$, whose value is $\frac{\epsilon^{ax}}{a}$. Thus, by successive application of equation (1) we have

$$\int x^n \epsilon^{ax}\, dx = \frac{\epsilon^{ax}}{a}\left[x^n - \frac{n}{a}x^{n-1} \ldots + (-1)^n \frac{n(n-1)\cdots 1}{a^n}\right].$$

Reduction of $\int \sin^m \theta\, d\theta$ *and* $\int \cos^m \theta\, d\theta$.

66. To obtain a formula of reduction, it is sometimes necessary to make a further transformation of the equation obtained by the method of parts. Thus, for the integral

$$\int \sin^m \theta\, d\theta,$$

the method of parts gives

$$\int \sin^m \theta\, d\theta = \sin^{m-1}\theta(-\cos\theta) + (m-1)\int \sin^{m-2}\theta \cos^2\theta\, d\theta.$$

Substituting in the latter integral $1 - \sin^2\theta$ for $\cos^2\theta$,

$$\int \sin^m\theta\,d\theta = -\sin^{m-1}\theta\cos\theta$$

$$+ (m-1)\int \sin^{m-2}\theta\,d\theta - (m-1)\int \sin^m\theta\,d\theta;$$

transposing and dividing, we have

$$\int \sin^m\theta\,d\theta = -\frac{\sin^{m-1}\theta\cos\theta}{m} + \frac{m-1}{m}\int \sin^{m-2}\theta\,d\theta, \quad . \quad . \quad . \quad (1)$$

a formula of reduction in which the exponent of $\sin\theta$ is diminished two units. By successive application of this formula, we have, for example:

$$\int \sin^6\theta\,d\theta = -\frac{\sin^5\theta\cos\theta}{6} + \frac{5}{6}\int \sin^4\theta\,d\theta$$

$$= -\frac{\sin^5\theta\cos\theta}{6} - \frac{5}{6}\frac{\sin^3\theta\cos\theta}{4} + \frac{5}{6}\frac{3}{4}\int \sin^2\theta\,d\theta$$

$$= -\frac{\sin^5\theta\cos\theta}{6} - \frac{5\sin^3\theta\cos\theta}{6\cdot4} - \frac{5\cdot3\sin\theta\cos\theta}{6\cdot4\cdot2} + \frac{5\cdot3\cdot1}{6\cdot4\cdot2}\theta.$$

67. By a process similar to that employed in deriving equation (1), or simply by putting $\theta = \frac{1}{2}\pi - \theta'$ in that equation, we find

$$\int \cos^m\theta\,d\theta = \frac{\cos^{m-1}\theta\sin\theta}{m} + \frac{m-1}{m}\int \cos^{m-2}\theta\,d\theta, \quad . \quad . \quad (2)$$

a formula of reduction, when m is positive.

68. It should be noticed that, when m is negative, equation (1) Art. 66 is not a formula of reduction, because the exponent in the new integral is in that case numerically greater than the exponent in the given integral. But, if we now regard the integral in the second member as the given one, the equation is readily converted into a formula of reduction. Thus, putting $-n$ for the negative exponent $m-2$, whence

$$m = -n + 2,$$

transposing and dividing, equation (1) becomes

$$\int \frac{d\theta}{\sin^n \theta} = -\frac{\cos\theta}{(n-1)\sin^{n-1}\theta} + \frac{n-2}{n-1}\int \frac{d\theta}{\sin^{n-2}\theta}. \quad . \quad . \quad . \quad (3)$$

Again, putting $\theta = \frac{1}{2}\pi - \theta'$ in this equation, we obtain

$$\int \frac{d\theta}{\cos^n \theta} = \frac{\sin\theta}{(n-1)\cos^{n-1}\theta} + \frac{n-2}{n-1}\int \frac{d\theta}{\cos^{n-2}\theta} \quad . \quad . \quad . \quad . \quad (4)$$

Reduction of $\int \sin^m\theta \cos^n\theta\, d\theta$.

69. If we put $dv = \sin^m\theta \cos\theta\, d\theta$, we have

$$\int \sin^m\theta \cos^n\theta\, d\theta = \frac{\cos^{n-1}\theta \sin^{m+1}\theta}{m+1}$$

$$+ \frac{n-1}{m+1}\int \sin^{m+2}\theta \cos^{n-2}\theta\, d\theta; \quad . \quad . \quad . \quad (1)$$

but, if in the same integral we put $dv = \cos^n\theta \sin\theta\, d\theta$, we have

$$\int \sin^m\theta \cos^n\theta\, d\theta = -\frac{\sin^{m-1}\theta \cos^{n+1}\theta}{n+1}$$

$$+ \frac{m-1}{n+1}\int \sin^{m-2}\theta \cos^{n+2}\theta\, d\theta. \quad . \quad . \quad . \quad (2)$$

When m and n are both positive, equation (1) is not a formula of reduction, since in the new integral the exponent of $\sin\theta$ is increased, while that of $\cos\theta$ is diminished. We therefore substitute in this integral

$$\sin^{m+2}\theta = \sin^m\theta\,(1 - \cos^2\theta),$$

so that the last term of the equation becomes

$$\frac{n-1}{m+1}\int \sin^m\theta \cos^{n-2}\theta\, d\theta - \frac{n-1}{m+1}\int \sin^m\theta \cos^n\theta\, d\theta.$$

Hence, by this transformation, the original integral is reproduced, and equation (1) becomes

$$\left[1 + \frac{n-1}{m+1}\right]\int \sin^m\theta \cos^n\theta\, d\theta = \frac{\sin^{m+1}\theta \cos^{n-1}\theta}{m+1}$$

$$+ \frac{n-1}{m+1}\int \sin^m\theta \cos^{n-2}\theta\, d\theta.$$

Dividing by $1 + \dfrac{n-1}{m+1} = \dfrac{m+n}{m+1}$, we have

$$\int \sin^m\theta \cos^n\theta\, d\theta = \frac{\sin^{m+1}\theta \cos^{n-1}\theta}{m+n}$$

$$+ \frac{n-1}{m+n}\int \sin^m\theta \cos^{n-2}\theta\, d\theta, \quad . \quad . \quad . \quad (3)$$

a formula of reduction by which the exponent of $\cos\theta$ is diminished two units.

By a similar process, from equation (2), or simply by putting $\theta = \frac{1}{2}\pi - \theta'$ in equation (3), and interchanging m and n, we obtain

$$\int \sin^m \theta \cos^n \theta \, d\theta = -\frac{\sin^{m-1}\theta \cos^{n+1}\theta}{m+n}$$

$$+ \frac{m-1}{m+n}\int \sin^{m-2}\theta \cos^n \theta \, d\theta, \quad . \quad . \quad . \quad (4)$$

a formula by which the exponent of $\sin\theta$ is diminished two units.

70. When n is positive and m negative, equation (1) of the preceding article is itself a formula of reduction, for both exponents are in that case numerically diminished. Putting $-m$ in place of m, the equation becomes

$$\int \frac{\cos^n \theta}{\sin^m \theta} d\theta = -\frac{\cos^{n-1}\theta}{(m-1)\sin^{m-1}\theta} - \frac{n-1}{m-1}\int \frac{\cos^{n-2}}{\sin^{m-2}} d\theta. \quad . \quad . \quad . \quad (5)$$

Similarly, when m is positive and n negative, equation (2) gives

$$\int \frac{\sin^m \theta}{\cos^n \theta} d\theta = \frac{\sin^{m-1}\theta}{(n-1)\cos^{n-1}\theta} - \frac{m-1}{n-1}\int \frac{\sin^{m-2}\theta}{\cos^{n-2}\theta} d\theta. \quad . \quad . \quad . \quad (6)$$

71. When m and n are both negative, putting $-m$ and $-n$ in place of m and n, equation (3) Art. 69 becomes

$$\int \frac{d\theta}{\sin^m \theta \cos^n \theta} = -\frac{1}{(m+n)\sin^{m-1}\theta \cos^{n+1}\theta}$$

$$+ \frac{n+1}{m+n}\int \frac{d\theta}{\sin^m \theta \cos^{n+2}\theta},$$

in which the exponent of $\cos\theta$ is numerically increased. We may therefore regard the integral in the second member as the integral to be reduced. Thus, putting n in place of $n + 2$, we derive

$$\int \frac{d\theta}{\sin^m \theta \cos^n \theta} = \frac{1}{(n-1)\sin^{m-1}\theta \cos^{n-1}\theta} + \frac{m+n-2}{n-1} \int \frac{d\theta}{\sin^m \theta \cos^{n-2}\theta}. \quad . \quad . \quad . \quad . \quad (7)$$

Putting $\theta = \frac{1}{2}\pi - \theta'$, and interchanging m and n, we have

$$\int \frac{d\theta}{\sin^m \theta \cos^n \theta} = -\frac{1}{(m-1)\sin^{m-1}\theta \cos^{n-1}\theta} + \frac{m+n-2}{m-1} \int \frac{d\theta}{\sin^{m-2} \theta \cos^{n}\theta}. \quad . \quad . \quad . \quad . \quad (8)$$

72. The application of the formulas derived in the preceding articles to definite integrals will be given in the next section. When the value of the indefinite integral is required, it should first be ascertained whether the given integral belongs to one of the directly integrable cases mentioned in Arts. 27 and 28. If it does not, the formulas of reduction must be used, and if m and n are integers, we shall finally arrive at a directly integrable form.

As an illustration, let us take the integral

$$\int \sin^2 \theta \cos^4 \theta \, d\theta.$$

Employing formula (4) Art. 69, by which the exponent of $\sin \theta$ is diminished, we have

$$\int \sin^2 \theta \cos^4 \theta \, d\theta = -\frac{\sin \theta \cos^5 \theta}{6} + \frac{1}{6} \int \cos^4 \theta \, d\theta.$$

The last integral can be reduced by means of formula (2) Art. 67, which, when $m = 4$, gives

$$\int \cos^4\theta\, d\theta = \frac{\cos^3\theta \sin\theta}{4} + \frac{3}{4}\int \cos^2\theta\, d\theta\,;$$

therefore

$$\int \sin^2\theta \cos^4\theta\, d\theta = \frac{\sin\theta\cos^5\theta}{6} + \frac{\cos^3\theta\sin\theta}{24} + \frac{\sin\theta\cos\theta}{16} + \frac{\theta}{16}.$$

73. Again, let the given integral be

$$\int \frac{\cos^6\theta\, d\theta}{\sin^3\theta}.$$

By equation (5), Art. 70, we have

$$\int \frac{\cos^6\theta\, d\theta}{\sin^3\theta} = -\frac{\cos^5\theta}{2\sin^2\theta} - \frac{5}{2}\int \frac{\cos^4\theta\, d\theta}{\sin\theta}.$$

We cannot apply the same formula to the new integral, since the denominator $m - 1$ vanishes; but putting $n = 4$ and $m = -1$, in equation (3) Art. 69, we have

$$\int \frac{\cos^4\theta\, d\theta}{\sin\theta} = \frac{\cos^3\theta}{3} + \int \frac{\cos^2\theta\, d\theta}{\sin\theta}$$

$$= \frac{\cos^3\theta}{3} + \int \frac{d\theta}{\sin\theta} - \int \sin\theta\, d\theta$$

$$= \frac{\cos^3\theta}{3} + \log\tan\frac{1}{2}\theta + \cos\theta.$$

Hence

$$\int \frac{\cos^6\theta\, d\theta}{\sin^3\theta} = -\frac{\cos^5\theta}{2\sin^2\theta} - \frac{5\cos^3\theta}{6} - \frac{5}{2}\log\tan\frac{1}{2}\theta - \frac{5}{2}\cos\theta.$$

Extension of the Formula.

74. Let

$$\int \phi(x)\,dx = \phi_{,}(x),$$

$$\int \phi_{,}(x)\,dx = \phi_{,,}(x),$$

etc., etc.;

then, if the functions $\phi_{,}(x), \phi_{,,}(x), \ldots\ \phi_{n}(x)$, which may be called the *successive integrals* of $\phi(x)$, are known, and also the successive derivatives of $f(x)$, we shall have

$$\int f(x)\,\phi(x)\,dx = f(x)\,\phi_{,}(x) - \int f'(x)\,\phi_{,}(x)\,dx$$

$$= f(x)\,\phi_{,}(x) - f'(x)\,\phi_{,,}(x) + \int f''(x)\,\phi_{,,}(x)\,dx.$$

Continuing this process, and writing for shortness $f, \phi_{,}, \ldots$ for $f(x), \phi_{,}(x) \ldots$ we have

$$\int f(x)\,\phi(x)\,dx = f\cdot\phi_{,} - f'\cdot\phi_{,,} + \cdots\cdots + (-1)^{n-1} f^{n-1}\cdot\phi_{n}$$

$$+ (-1)^{n}\int f^{n}\cdot\phi_{n}\cdot dx.$$

The application of this formula is equivalent to the use of a formula of reduction. Thus the value of $\int x^{n}\,\varepsilon^{ax}$ given in Art. 65, may be derived immediately from it.

Taylor's Theorem.

75. If, in the formula of the preceding article, we put

$$f(x) = F'(x_0 + h - x), \qquad \text{and} \qquad \phi(x) = 1,$$

x_0 and h being constants,

$$f'(x) = -F''(x_0 + h - x), \qquad f''(x) = F'''(x_0 + h - x), \text{ etc.};$$

and $\quad \phi_{,}(x) = x, \quad \phi_{,,}(x) = \dfrac{x^2}{1 \cdot 2}, \quad \phi_{,,,}(x) = \dfrac{x^3}{1 \cdot 2 \cdot 3}$, etc.

Hence

$$\int F'(x_0 + h - x)\,dx = F'(x_0 + h - x) \cdot x + F''(x_0 + h - x)\frac{x^2}{1 \cdot 2}$$

$$+ \cdots\cdots + \int F^{n+1}(x_0 + h - x)\frac{x^n}{1 \cdot 2 \cdots n}\,dx.$$

Now $$\int F'(x_0 + h - x)\,dx = -F(x_0 + h - x);$$

hence, applying the limits 0 and h, we have

$$F(x_0 + h) = F(x_0) + F'(x_0)\,h + F''(x_0)\frac{h^2}{1 \cdot 2} + \cdots\cdots$$

$$+ \int_0^h F^{n+1}(x_0 + h - x)\frac{x^n\,dx}{1 \cdot 2 \cdots n}.$$

This formula is Taylor's Theorem, with the remainder expressed in the form of a definite integral.

Examples VI.

1. $\int_0^1 \sin^{-1} x \, dx,$ $\quad x \sin^{-1} x + \sqrt{(1 - x^2)}\Big]_0^1 = \frac{\pi}{2} - 1.$

2. $\int \sec^{-1} x \, dx,$ $\quad x \sec^{-1} x - \log [x + \sqrt{(x^2 - 1)}].$

3. $\int_0^1 \tan^{-1} x \, dx,$ $\quad \frac{\pi}{4} - \frac{\log 2}{2}.$

4. $\int x^n \log x \, dx,$ $\quad \frac{x^{n+1}}{n+1}\left[\log x - \frac{1}{n+1}\right].$

5. $\int_0^{\frac{\pi}{2}} \theta \sin \theta \, d\theta,$ $\quad 1.$

6. $\int_0^{\frac{\pi}{2}} \theta \cos m\theta \, d\theta,$ $\quad \frac{\pi}{2m} - \frac{1}{m^2}.$

7. $\int x \tan^{-1} x \, dx,$ $\quad \frac{1 + x^2}{2} \tan^{-1} x - \frac{x}{2}.$

8. $\int_0 x^2 \varepsilon^x \, dx,$ $\quad x^2 \varepsilon^x - 2x\varepsilon^x + 2\varepsilon^x - 2.$

9. $\int x \sec^{-1} x \, dx,$ $\quad \frac{1}{2}[x^2 \sec^{-1} x - \sqrt{(x^2 - 1)}].$

10. $\int_0^{\frac{\pi}{2}} \theta \sin\left[\frac{\pi}{4} + \theta\right] d\theta,$ $\quad -\theta \cos\left(\frac{\pi}{4} + \theta\right) + \sin\left(\frac{\pi}{4} + \theta\right)\Big]_0^{\frac{\pi}{2}} = \frac{\pi\sqrt{2}}{4}.$

11. $\int x \sec^2 x \, dx,$ $\qquad x \tan x + \log \cos x.$

12. $\int x \tan^2 x \, dx \left[= \int x \, (\sec^2 x - 1) \, dx \right],$ $\qquad x \tan x + \log \cos x - \frac{1}{2} x^2.$

13. $\int x^2 \sin x \, dx,$ $\qquad 2x \sin x + 2 \cos x - x^2 \cos x.$

14. $\int_0^1 x \sin^{-1} x \, dx,$ $\qquad \left. \frac{1}{2} x^2 \sin^{-1} x \right]_0^1 - \frac{1}{2} \int_0^{\frac{\pi}{2}} \sin^2 \theta \, d\theta = \frac{\pi}{8}.$

15. $\int x^2 \tan^{-1} x \, dx,$ $\qquad \frac{x^3 \tan^{-1} x}{3} - \frac{x^2}{6} + \frac{\log (1 + x^2)}{6}.$

16. $\int_0^1 x^2 \sin^{-1} x \, dx,$ $\qquad \left. \frac{1}{3} x^3 \sin^{-1} x + \frac{2 + x^2}{9} \sqrt{(1 - x^2)} \right]_0^1 = \frac{\pi}{6} - \frac{2}{9}$

17. $\int_0^\infty \varepsilon^{-x} \cos x \, dx,$ $\qquad \left. \frac{\varepsilon^{-x} (\sin x - \cos x)}{2} \right]_0^\infty = \frac{1}{2}.$

18. $\int \varepsilon^{x \tan \beta} \cos x \, dx,$ $\qquad \cos \beta \, \varepsilon^{x \tan \beta} \sin (\beta + x).$

19. $\int \varepsilon^{-x} \sin^2 x \, dx \left[= \frac{1}{2} \int \varepsilon^{-x} (1 - \cos 2x) \, dx \right],$

$$\frac{\varepsilon^{-x}}{10} (\cos 2x - 2 \sin 2x - 5).$$

20. $\int_0^{\frac{\pi}{4}} \varepsilon^{\theta} \sin \theta \, d\theta,$ $\qquad \left. \frac{\varepsilon^{\theta}}{2} (\sin \theta - \cos \theta) \right]_0^{\frac{\pi}{4}} = \frac{1}{2}.$

21. $\int \varepsilon^x \sin x \cos x \, dx,$ $\qquad \frac{\varepsilon^x}{10}(\sin 2x - 2\cos 2x).$

22. $\int_0 \sin^4 m\theta \, d\theta,$ $\qquad \frac{-\sin^3 m\theta \cos m\theta}{4m} + \frac{3\theta}{8} - \frac{3 \sin m\theta \cos m\theta}{8m}.$

23. Derive a formula of reduction for $\int (\log x)^n x^m \, dx$, and deduce from it the value of $\int (\log x)^3 x^2 \, dx$.

$$\int (\log x)^n x^m \, dx = (\log x)^n \frac{x^{m+1}}{m+1} - \frac{n}{m+1} \int (\log x)^{n-1} x^m \, dx.$$

$$\int (\log x)^3 x^2 \, dx = (\log x)^3 \frac{x^3}{3} - (\log x)^2 \frac{x^3}{3} + \frac{2x^3 \log x}{9} - \frac{2x^3}{27}.$$

24. $\int x \cos^2 x \, dx,$ $\qquad \frac{1}{2} x \sin x \cos x - \frac{1}{4} \sin^2 x + \frac{1}{4} x^2.$

25. $\int_1 x^2 \sec^{-1} x \, dx,$ $\qquad \frac{x^3 \sec^{-1} x}{3} - \frac{x\sqrt{(x^2-1)}}{6} - \frac{\log [x + \sqrt{(x^2-1)}]}{6}.$

26. Derive a formula of reduction for $\int x^n \sin(x+\alpha) \, dx$, and deduce from it the value of $\int x^5 \cos x$.

$$\int x^n \sin(x+\alpha)\, dx = -x^n \sin\left[x + \alpha + \frac{\pi}{2}\right]$$

$$+ n \int x^{n-1} \sin\left[x + \alpha + \frac{\pi}{2}\right] dx.$$

$$\int x^5 \cos x \, dx = (x^5 - 20x^3 + 120x) \sin x + (5x^4 - 60x^2 + 120) \cos x.$$

27. $\int \cos^2\theta \sin^4\theta \, d\theta$, $\quad \frac{\sin^5\theta\cos\theta}{6} - \frac{\sin^3\theta\cos\theta}{24} + \frac{1}{16}[\theta - \sin\theta\cos\theta]$.

28. $\int_0^{\frac{\pi}{4}} \cos^4\theta \sin^4\theta \, d\theta$, $\quad \frac{1}{32}\int_0^{\frac{\pi}{2}} \sin^4\theta' \, d\theta' = \frac{\pi}{512}$.

29. $\int_0^{\frac{\pi}{4}} \cos^4\theta \, d\theta$, $\quad \left.\frac{\sin\theta\cos^3\theta}{4} + \frac{3\sin\theta\cos\theta}{8} + \frac{3\theta}{8}\right]_0^{\frac{\pi}{4}} = \frac{8+3\pi}{32}$.

30. $\int_0^{\frac{\pi}{3}} \cos^6\theta \, d\theta$,

$$\left.\frac{\sin\theta\cos\theta\,(8\cos^4\theta + 10\cos^2\theta + 15) + 15\theta}{48}\right]_0^{\frac{\pi}{3}} = \frac{9\sqrt{3}+10\pi}{96}.$$

31. $\int \frac{\cos^4\theta}{\sin^3\theta} d\theta$, $\quad -\frac{\cos^3\theta}{2\sin^2\theta} - \frac{3\cos\theta}{2} - \frac{3\log\tan\frac{1}{2}\theta}{2}$.

32. $\int \frac{\sin^2\theta}{\cos^5\theta} d\theta$, $\quad \frac{\sin\theta}{4\cos^4\theta} - \frac{\sin\theta}{8\cos^2\theta} - \frac{1}{8}\log\tan\left[\frac{\pi}{4} + \frac{\theta}{2}\right]$.

33. $\int \frac{\sin^6\theta}{\cos^4\theta} d\theta$, $\quad \frac{\sin^5\theta}{3\cos^3\theta} - \frac{5\sin^3\theta}{3\cos\theta} + \frac{5}{2}[\theta - \sin\theta\cos\theta]$.

34. $\int_{\frac{\pi}{4}}^{\frac{\pi}{2}} \frac{\cos^6\theta}{\sin^2\theta} d\theta$, $\quad -\left.\frac{\cos^5\theta}{\sin\theta}\right]_{\frac{\pi}{4}}^{\frac{\pi}{2}} - 5\int_{\frac{\pi}{4}}^{\frac{\pi}{2}} \cos^4\theta \, d\theta = \frac{48 - 15\pi}{32}$.

35. $\int_0^{\frac{\pi}{2}} \frac{d\theta}{(1+\cos\theta)^2}$ $\quad \frac{1}{2}\int_0^{\frac{\pi}{4}} \frac{d\theta'}{\cos^4\theta'} = \frac{2}{3}$.

36. $\int \frac{d\theta}{\sin\theta\cos^4\theta}$, $\quad \frac{1}{3\cos^3\theta} + \frac{1}{\cos\theta} + \log\tan\frac{\theta}{2}$.

37. $\displaystyle\int\frac{d\theta}{\sin\theta\sin^2 2\theta},\qquad -\frac{1}{4\sin^2\theta\cos\theta}-\frac{3\cos\theta}{8\sin^2\theta}+\frac{3}{8}\log\tan\frac{\theta}{2}.$

38. Prove that when n is odd

$$\int\frac{d\theta}{\sin\theta\cos^n\theta}=\frac{\sec^{n-1}\theta}{n-1}+\frac{\sec^{n-3}\theta}{n-3}+\cdots\cdots+\log\tan\theta;$$

and when n is even

$$\int\frac{d\theta}{\sin\theta\cos^4\theta}=\frac{\sec^{n-1}\theta}{n-1}+\frac{\sec^{n-3}\theta}{n-3}+\cdots\cdots+\log\tan\frac{\theta}{2}.$$

39. $\displaystyle\int\frac{d\theta}{\sin^3\theta\cos^4\theta},\quad -\frac{1}{2\sin^2\theta\cos^3\theta}+\frac{5}{2}\left[\frac{\sec^3\theta}{3}+\sec\theta+\log\tan\frac{\theta}{2}\right].$

40. $\displaystyle\int\frac{x^2\,dx}{\sqrt{(x^2-1)}},\quad$ *Put* $x=\sec\theta.$

$$\tfrac{1}{2}x\sqrt{(x^2-1)}+\tfrac{1}{2}\log[x+\sqrt{(x^2-1)}].$$

41. $\displaystyle\int(a^2-x^2)^{\frac{3}{2}}\,dx,\qquad \frac{(5a^2-2x^2)\,x\sqrt{(a^2-x^2)}}{8}+\frac{3a^4}{8}\sin^{-1}\frac{x}{a}.$

42. $\displaystyle\int_0^a\frac{dx}{(a^2+x^2)^3},\qquad \frac{1}{a^5}\int_0^{\frac{\pi}{4}}\cos^4\theta\,d\theta=\frac{3\pi+8}{32a^5}.$

43. $\displaystyle\int(a^2+x^2)\sqrt{(a^2-x^2)}\,dx,\quad \frac{5a^4}{8}\sin^{-1}\frac{x}{a}+\frac{x\sqrt{(a^2-x^2)}[3a^2+2x^2]}{8}.$

44. $\displaystyle\int\frac{x^2\,dx}{(x^2-a^2)^{\frac{3}{2}}},\qquad -\frac{x}{\sqrt{(x^2-a^2)}}+\log[x+\sqrt{(x^2-a^2)}].$

45. $\displaystyle\int\frac{x^2\,dx}{(x^2+1)^3},\qquad \frac{x(x^2-1)}{8(x^2+1)^2}+\frac{\tan^{-1}x}{8}.$

$$\frac{\sin\theta\cos\theta}{\sin\theta+\cos\theta}.$$

47. Derive a formula for the reduction of $\int x\sec^n x\,dx$; and referring to Ex. 11, thence show that this is an integrable form when n is an even integer. Give the result when $n=4$.

$$\int x\sec^n x\,dx = \frac{x\sec^{n-2}x\tan x}{n-1} - \frac{\sec^{n-2}x}{(n-1)(n-2)}$$

$$+\frac{n-2}{n-1}\int x\sec^{n-2}x\,dx.$$

$$\int x\sec^4 x\,dx = \frac{x\sec^2 x\tan x}{3} - \frac{\sec^2 x}{6} + \frac{2}{3}[x\tan x + \log\cos x].$$

48. Derive a formula of reduction for $\int x\cos^n x\,dx$, and deduce from it the value of $\int x\cos^3 x\,dx$.

$$\int x\cos^n x\,dx = \frac{x\cos^{n-1}x\sin x}{n} + \frac{\cos^n x}{n^2} + \frac{n-1}{n}\int x\cos^{n-2}x\,dx.$$

$$\int x\cos^3 x\,dx = \frac{x\sin x}{3}(\cos^2 x+2) + \frac{\cos x}{9}(\cos^2 x+6).$$

49. Find the area between the curve

$$y=\sec^{-1}x,$$

the axis of x, and the ordinate corresponding to $x = 2$.

$$\frac{2\pi}{3} - \log[2 + \sqrt{3}] = 0.77744.$$

50. Find the area between the axis of x, the curve

$$y = \tan^{-1} x,$$

and the ordinate corresponding to $x = 1$. $\frac{\pi}{4} - \frac{\log 2}{2} = 0.43882.$

VII.

Definite Integrals.

76. Before proceeding to transformations of definite integrals involving the values of the limits, it is necessary to resume the consideration of the relations between a definite integral and its limits, as defined in the first section.

By definition, the symbol

$$\int_a^X f(x)\,dx$$

denotes the quantity generated at the rate

$$f(x)\frac{dx}{dt},$$

while x passes from the initial value a to the final value X. The rate of x is arbitrary, and may be assumed constant; but in that case its sign must be the same as that of the increment received by x; that is, the sign of dx is the same as that of $X - a$.

These considerations often serve to determine the sign of an integral. Thus

$$\int_0^\pi \frac{\sin x\, dx}{x}$$

denotes a positive quantity, because dx is positive, and $\dfrac{\sin x}{x}$ is positive for all values of x between 0 and π.

77. Now let $F(x)$ denote a value of the indefinite integral, so that

$$d\{F(x)\} = f(x)\, dx;$$

thus $f(x)$ is the derivative of $F(x)$. Then, *supposing* F (x) *to vary continuously as* x *passes from* a *to* X; that is, to have no infinite or imaginary values for values of x between a and X, the integral is the actual increment received by $F(x)$, while x passes from a to X. In this case, therefore

$$\int_a^X f(x)\, dx = F(X) - F(a) \quad . \quad . \quad . \quad . \quad . \quad (1).$$

If, on the other hand, there is any value, α, between a and X, such that

$$F(\alpha) = \infty,$$

equation (1) does not hold true. For example,

$$\int \frac{dx}{x^2} = -\frac{1}{x},$$

and in the case of the definite integral

$$\int_{-1}^1 \frac{dx}{x^2}$$

x passes through the value zero, for which $F(x)$ is infinite; *we cannot therefore write*

$$\int_{-1}^{1} \frac{dx}{x^2} = -\frac{1}{x}\Big]_{-1}^{1} = -2.$$

This result indeed is obviously false, since dx is here positive, and x^2 is never negative for real values of x. The value of the integral is in fact infinite, since the increments received by $-\frac{1}{x}$, while x passes from -1 to 0, and while x passes from 0 to 1, are both infinite and positive.

78. Since the derivative of a function becomes infinite when the function becomes infinite [Diff. Calc., Art. 104; Abridged Ed., Art. 89], we can have $F(a) = \infty$ only when $f(a) = \infty$; but it is to be noticed that $F(x)$ does not necessarily become infinite when $f(x)$ becomes infinite. Thus, in

$$\int_{-1}^{8} \frac{dx}{x^{\frac{1}{3}}}$$

$f(x) = x^{-\frac{1}{3}}$, which becomes infinite for $x = 0$, a value of x between the limits; but since

$$\int x^{-\frac{1}{3}}\, dx = \tfrac{3}{2} x^{\frac{2}{3}}$$

the indefinite integral $F(x)$ does not become infinite. Therefore equation (1) holds true, and

$$\int_{-1}^{8} \frac{dx}{x^{\frac{1}{3}}} = \frac{3}{2} x^{\frac{2}{3}}\Big]_{-1}^{8} = \frac{9}{2}.$$

79. We have, in the preceding articles, assumed that the independent variable varies uniformly in passing from the lower to the upper limit; but when a change of independent variable is made, the new variable does not generally vary uniformly between its limits. It is, however, obvious, that, in equation (1), Art. 77, x may vary in any manner whatever in passing from a to X, provided that $F(x)$ *remains throughout a continuous one-valued function;* x may even pass through infinity, provided $F(x)$ is finite and one-valued when $x = \infty$.

Multiple-Valued Integrals.

80. When the indefinite integral is a multiple-valued function, a particular value of this function must of course be employed, and it is necessary to take care that this value varies continuously while x passes from the lower to the upper limit. In the fundamental formula (j) it is sufficient (provided the radical $\sqrt{(1 - x^2)}$ does not change sign), to limit the meaning of the symbols $\sin^{-1} x$ and $\cos^{-1} x$ to the primary values of these symbols (see Diff. Calc., Arts. 54 and 55), since these values are so taken as to vary continuously while x passes through all its possible values from -1 to $+1$.

81. In the case of formula (k) the primary value of $\tan^{-1} x$ is so defined that, as x passes from $-\infty$ to $+\infty$, the primary value varies continuously from $-\frac{1}{2}\pi$ to $+\frac{1}{2}\pi$. We may therefore employ the primary value at both limits, *unless* x *passes through infinity*, as in the following example. Given the integral

$$\int_0^{\frac{5\pi}{6}} \frac{d\theta}{\cos^2\theta + 9\sin^2\theta} = \int_0^{\frac{5\pi}{6}} \frac{\sec^2\theta\, d\theta}{1 + 9\tan^2\theta},$$

if we put $\tan\theta = x$, this becomes

$$\int_0^{-\frac{\sqrt{3}}{3}} \frac{dx}{1 + 9x^2} = \frac{1}{3}\tan^{-1} 3x\Big]_0^{-\frac{\sqrt{3}}{3}} = \frac{1}{3}\left[\tan^{-1}(-\sqrt{3}) - \tan^{-1} 0\right].$$

But here it is to be noticed, that, as θ passes from 0 to $\frac{5}{6}\pi$, x

passes through infinity when $\theta = \frac{1}{2}\pi$. Hence, if the value of $\tan^{-1} 3x$ is taken as 0 at the lower limit, it is to be regarded as increasing and passing through $\frac{1}{2}\pi$, when $x = \infty$, so that its value at the upper limit is $\frac{2}{3}\pi$, and not $-\frac{1}{3}\pi$. Hence

$$\int_0^{\frac{5\pi}{6}} \frac{d\theta}{\cos^2\theta + 9\sin^2\theta} = \frac{2\pi}{9}.$$

82. When the symbol $\cot^{-1} x$ is employed, the primary value, defined in the same manner as in the case of $\tan^{-1} x$, cannot be taken at both limits *when x passes through zero*. Thus, using the second form of (k), Art. 10, we have

$$\int_1^{-1} \frac{dx}{1 + x^2} = \cot^{-1} 1 - \cot^{-1}(-1),$$

in which, if $\cot^{-1} 1$ is taken as $\frac{1}{4}\pi$, $\cot^{-1}(-1)$ must be taken as $\frac{3}{4}\pi$. Thus

$$\int_1^{-1} \frac{dx}{1 + x^2} = -\frac{1}{2}\pi.$$

Formulas of Reduction for Definite Integrals.

83. The limits of a definite integral are very often such as to simplify materially the formula of reduction appropriate to it. For example, to reduce

$$\int_0^\infty x^n \epsilon^{-x} dx.$$

we have by the method of parts

$$\int x^n \epsilon^{-x} dx = -x^n \epsilon^{-x} + n\int \epsilon^{-x} x^{n-1} dx.$$

Now, supposing n positive, the quantity $x^n \varepsilon^{-x}$ vanishes when $x=0$, and also when $x=\infty$ [See Diff. Calc., Art. 107; Abridged Ed., Art. 91]. Hence, applying the limits 0 and ∞,

$$\int_0^\infty x^n \varepsilon^{-x}\,dx = n\int_0^\infty x^{n-1}\varepsilon^{-x}\,dx.$$

By successive application of this formula we have, when n is an integer,

$$\int_0^\infty x^n \varepsilon^{-x}\,dx = n(n-1)\cdot\cdots\cdots\cdot 2\cdot 1.$$

84. From equation (1) Art. 66, supposing $m > 1$, we have

$$\int_0^{\frac{\pi}{2}} \sin^m\theta\,d\theta = \frac{m-1}{m}\int_0^{\frac{\pi}{2}} \sin^{m-2}\theta\,d\theta.$$

If m is an integer, we shall, by successive application of this formula, finally arrive at $\int_0^{\frac{\pi}{2}} d\theta = \frac{\pi}{2}$ or $\int_0^{\frac{\pi}{2}} \sin\theta\,d\theta = 1$, according as m is even or odd. Hence

if m is even, $$\int_0^{\frac{\pi}{2}} \sin^m\theta\,d\theta = \frac{(m-1)(m-3)\cdots\cdots 1}{m(m-2)\cdots\cdots 2}\cdot\frac{\pi}{2}, \quad\ldots (P)$$

and if m is odd, $$\int_0^{\frac{\pi}{2}} \sin^m\theta\,d\theta = \frac{(m-1)(m-3)\cdots\cdots 2}{m(m-2)\cdots\cdots 1}. \quad\ldots (P')$$

85. From equations (3) and (4) Art. 69, we derive

$$\int_0^{\frac{\pi}{2}} \sin^m \theta \cos^n \theta \, d\theta = \frac{n-1}{m+n} \int_0^{\frac{\pi}{2}} \sin^m \theta \cos^{n-2} \theta \, d\theta,$$

and
$$\int_0^{\frac{\pi}{2}} \sin^m \theta \cos^n \theta \, d\theta = \frac{m-1}{m+n} \int_0^{\frac{\pi}{2}} \sin^{m-2} \theta \cos^n \theta \, d\theta.$$

By successive application of these formulas, we shall have for the final integral one of the four forms

$$\int_0^{\frac{\pi}{2}} d\theta, \quad \int_0^{\frac{\pi}{2}} \sin \theta \, d\theta, \quad \int_0^{\frac{\pi}{2}} \cos \theta \, d\theta, \quad \text{or} \quad \int_0^{\frac{\pi}{2}} \sin \theta \cos \theta \, d\theta.$$

The numerator of the final fraction $\left(\frac{n-1}{m+n} \text{ or } \frac{m-1}{m+n}\right)$ is in each case either 2 or 1. In the first case, the value of the final integral is $\frac{1}{2}\pi$, and the final denominator is 2: in the second and third cases, the value of the final integral is 1, and the final denominator is 3: in the fourth case, the value of the final integral is $\frac{1}{2}$, and the final denominator is 4. Therefore (since the factors in the denominator proceed by intervals of 2), it is readily seen that we may write

$$\int_0^{\frac{\pi}{2}} \sin^m \theta \cos^n \theta \, d\theta = \frac{(m-1)(m-3)\cdots\cdot(n-1)(n-3)\cdots}{(m+n)(m+n-2)\cdots\cdots\cdots} \alpha, \quad (Q)$$

provided that each series of factors is carried to 2 or 1, *and* α *is taken equal to unity, except when* m *and* n *are both even, in which case* $\alpha = \frac{1}{2}\pi$.

86. The following propositions are obvious consequences of equation (1), Art. 77.

$$\int_a^b f(x)\,dx = -\int_b^a f(x)\,dx \quad . \quad . \quad . \quad . \quad . \quad . \quad . \quad (1)$$

$$\int_a^b f(x)\,dx = \int_a^c f(x)\,dx + \int_c^b f(x)\,dx \quad . \quad . \quad . \quad (2)$$

Again, if we put $x = a + b - z$, we have

$$\int_a^b f(x)\,dx = -\int_b^a f(a+b-z)\,dz = \int_a^b f(a+b-z)\,dz$$

by (1), or since it is indifferent whether we write z or x for the variable in a definite integral,

$$\int_a^b f(x)\,dx = \int_a^b f(a+b-x)\,dx \quad . \quad . \quad . \quad . \quad (3)$$

If $a = 0$, we have the particular case

$$\int_0^b f(x)\,dx = \int_0^b f(b-x)\,dx \quad . \quad . \quad . \quad . \quad (4)$$

87. As an application of formula (4), we have

$$\int_0^{\frac{\pi}{2}} \cos^m \theta \, d\theta = \int_0^{\frac{\pi}{2}} \cos^m\left(\frac{\pi}{2} - \theta\right) d\theta = \int_0^{\frac{\pi}{2}} \sin^m \theta \, d\theta \quad . \quad . \quad . \quad . \quad (1)$$

Hence the value of $\int_0^{\frac{\pi}{2}} \cos^m \theta \, d\theta$ as well as that of $\int_0^{\frac{\pi}{2}} \sin^m \theta \, d\theta$ is given by formulas (P) and (P'). The values of these integrals are readily found when the limits are any multiples of $\frac{1}{2}\pi$. For, by equation (2) of the preceding article, we may sum the values in the several quadrants. But, putting $\theta = k\frac{\pi}{2} + \theta'$, and employing equation (1), we have

$$\int_{k\frac{\pi}{2}}^{(k+1)\frac{\pi}{2}} \sin^m \theta \, d\theta = \pm \int_{k\frac{\pi}{2}}^{(k+1)\frac{\pi}{2}} \cos^m \theta \, d\theta = \pm \int_0^{\frac{\pi}{2}} \sin^m \theta \, d\theta, \quad . \quad . \quad . \quad (2)$$

in which the sign to be used is determined by that of $\sin^m \theta$ or $\cos^m \theta$ in the given quadrant.

In like manner the value of the integral in formula (Q) is numerically the same in every quadrant, and its sign is the same as that of $\sin^m \theta \cos^n \theta$ in the given quadrant.

Change of Independent Variable in a Definite Integral.

88. It is often useful to make such a change of independent variable as will leave unchanged, or simply interchange, the values of the limits. As an illustration, let us take the definite integral

$$u = \int_0^\infty \frac{\log x}{1 + x + x^2} dx.$$

If we put $x = \frac{1}{y}$, whence $\log x = -\log y$, and $dx = -\frac{dy}{y^2}$,

$$u = \int_\infty^0 \frac{\log y}{y^2 + y + 1}\,dy = -u;$$

whence we infer that

$$u = \int_0^\infty \frac{\log x}{1 + x + x^2}\,dx = 0.$$

89. Again, let

$$u = \int_0^\infty \frac{\log x}{a^2 + x^2}\,dx.$$

Putting $x = \frac{a^2}{y}$, we have

$$u = \int_0^\infty \frac{2\log a - \log y}{a^2 + y^2}\,dy = 2\log a \int_0^\infty \frac{dy}{a^2 + y^2} - u;$$

hence

$$\int_0^\infty \frac{\log x}{a^2 + x^2}\,dx = \frac{\pi \log a}{2a}.$$

Differentiation of an Integral.

90. The integral $\int_a^x f(x)\,dx$ is by definition a function of x, whose derivative, with reference to x, is $f(x)$. Thus, putting

$$U = \int_a^x f(x)\,dx,$$

$$\frac{dU}{dx} = f(x).$$

This gives the derivative of an integral with reference to its upper limit. By reversing the limits we have, in like manner,

$$\frac{dU}{da} = -f(a),$$

when the lower limit is regarded as variable.

91. Now writing the integral in the form

$$U = \int_a u\,dx, \quad . \quad . \quad . \quad . \quad . \quad . \quad . \quad . \quad (1)$$

if u is a function of some other quantity, α, independent of x and a, U is also a function of α, and therefore admits of a derivative with reference to α. From (1) we have

$$\frac{dU}{dx} = u,$$

whence

$$\frac{d}{d\alpha}\frac{dU}{dx} = \frac{du}{d\alpha}.$$

By the principle of differentiation with respect to independent variables [See Diff. Calc., Art. 401; Abridged Ed., Art. 200].

$$\frac{d}{dx}\frac{dU}{d\alpha} = \frac{d}{d\alpha}\frac{dU}{dx}.$$

Therefore

$$\frac{d}{dx}\frac{dU}{d\alpha} = \frac{du}{d\alpha};$$

and by integration

$$\frac{dU}{d\alpha} = \int\frac{du}{d\alpha}dx + C \quad . \quad . \quad . \quad . \quad . \quad . \quad . \quad . \quad (2)$$

Now, in equation (1), U is a function of x and α which, when $x = a$, is equal to zero, independently of the value of α. In other words, it is a constant with reference to α, when $x = a$; therefore $\frac{dU}{d\alpha} = 0$ when $x = a$. If, then, we use a as a lower limit in equation (2), we shall have $C = 0$. Therefore

$$\frac{dU}{d\alpha} = \int_a \frac{du}{d\alpha} dx \quad . \quad . \quad . \quad . \quad . \quad . \quad . \quad . \quad (3)$$

Substituting for x any value b independent of α, we have

$$\frac{d}{d\alpha}\int_a^b u\, dx = \int_a^b \frac{d}{d\alpha} u\, dx, \quad . \quad . \quad . \quad . \quad . \quad . \quad (4)$$

which expresses that *an integral may be differentiated with reference to a quantity of which the limits are independent, by differentiating the expression under the integral sign.*

92. By means of this theorem, we may derive from an integral whose value is known, the values of certain other integrals. Thus, from the first fundamental integral,

$$\int x^n\, dx = \frac{x^{n+1}}{n+1}, \quad . \quad . \quad . \quad . \quad . \quad . \quad . \quad . \quad (1)$$

we derive, by differentiating with reference to n,

$$\int x^n \log x\, dx = \frac{(n+1)x^{n+1}\log x - x^{n+1}}{(n+1)^2},$$

the result being the same as that which is obtained by the method of parts.

93. The principal application of this method, however, is to definite integrals, when the limits are such as materially to simplify the value of the original integral. Thus, equation (1) of the preceding article gives

$$\int_0^1 x^n\,dx = \frac{1}{n+1},$$

whence, by successive differentiation,

$$\int_0^1 x^n \log x\,dx = -\frac{1}{(n+1)^2},$$

$$\int_0^1 x^n (\log x)^2\,dx = \frac{1\cdot 2}{(n+1)^3},$$

. . . .

$$\int_0^1 x^n (\log x)^r\,dx = (-1)^r \frac{1\cdot 2 \cdots\cdots r}{(n+1)^{r+1}}.$$

Integration under the Integral Sign.

94. Let u be a function of x and α, and let a and α_0 be constants; then the integral

$$U = \int_{\alpha_0}\left[\int_a u\,dx\right] d\alpha, \quad \dots\dots \quad (1)$$

is a function of x and α, which vanishes when $\alpha = \alpha_0$, independently of the value of x, and when $x = a$, independently of the value of α. From (1)

$$\frac{dU}{d\alpha} = \int_a u\,dx, \qquad \text{whence} \qquad \frac{d}{dx}\frac{dU}{d\alpha} = u;$$

therefore $\quad \dfrac{d}{d\alpha}\dfrac{dU}{dx} = u, \qquad$ whence $\qquad \dfrac{dU}{dx} = \int u\,d\alpha + C.$

Now $\dfrac{dU}{dx}$ must vanish when $\alpha = \alpha_0$, since this supposition makes U independent of x; therefore, if we use α_0 for a lower limit in the last equation, we must have $C = 0$; therefore

$$\frac{dU}{dx} = \int_{\alpha_0} u \, d\alpha,$$

and since u vanishes when $x = a$,

$$U = \int_a \left[\int_{\alpha_0} u \, d\alpha\right] dx. \quad . \quad . \quad . \quad . \quad . \quad (2)$$

Comparing the values of U in equations (1) and (2), we have

$$\int_{\alpha_0} \int_a u \, dx \, d\alpha = \int_a \int_{\alpha_0} u \, d\alpha \, dx.$$

It is evident that we may also write

$$\int_{\alpha_0}^{\alpha_1} \int_a^b u \, dx \, d\alpha = \int_a^b \int_{\alpha_0}^{\alpha_1} u \, d\alpha \, dx, \quad . \quad . \quad . \quad . \quad . \quad (3)$$

provided that the limits of each integration are independent of the other variable.

95. By means of this formula, a new integral may be derived from the value of a given integral, provided we can integrate, with reference to the other variable, both the expressions under the integral sign and also the value of the integral. Thus, from

$$\int_0^1 x^n \, dx = \frac{1}{n+1},$$

by multiplying by dn, and integrating between the limits r and s, we derive

$$\int_0^1 \int_r^s x^n \, dn \, dx = \int_r^s \frac{dn}{n+1},$$

whence

$$\int_0^1 \frac{x^s - x^r}{\log x} \, dx = \log \frac{s+1}{r+1}.$$

96. When the derivative of a proposed integral with reference to α is a known integral, we can sometimes derive its value by integrating the latter with reference to α. Thus, let

$$u = \int_0^\infty \frac{\epsilon^{-\alpha x} - \epsilon^{-\beta x}}{x} \, dx. \qquad (1)$$

In this case

$$\frac{du}{d\alpha} = \int_0^\infty -\epsilon^{-\alpha x} \, dx = \left[\frac{\epsilon^{-\alpha x}}{\alpha}\right]_0^\infty = -\frac{1}{\alpha};$$

hence, integrating, $\quad u = -\log \alpha + C = \log \dfrac{\beta}{\alpha} \qquad (2)$

since in (1) u vanishes when $\alpha = \beta$.

The Definite Integral Regarded as the Limiting Value of a Sum.

97. Let A denote the greatest, and B the least value assumed by $f(x)$, while x varies from a to b. Then it is evident that

$$\int_a^b f(x) \, dx < \int_a^b A \, dx; \qquad (1)$$

for, while x passes from a to b, the rate of the former integral is generally less, and never greater than the rate of the latter. In like manner

$$\int_a^b f(x)\,dx > \int_a^b B\,dx. \quad . \quad . \quad . \quad . \quad . \quad . \quad . \quad (2)$$

The values of the integrals in the second members of equations (1) and (2) are $A\,(b-a)$ and $B\,(b-a)$ respectively. Therefore, if we assume

$$\int_a^b f(x)\,dx = M\,(b-a), \quad . \quad . \quad . \quad . \quad . \quad . \quad . \quad (3)$$

we shall have $$A > M > B.$$

The quantity M in equation (3) is called the *mean value* of the function $f(x)$ for the interval between a and b.

98. Let

$$b - a = n\,\Delta x; \quad . \quad . \quad . \quad . \quad . \quad . \quad . \quad . \quad (4)$$

then the $n+1$ values of x,

$$a, \qquad a + \Delta x, \qquad a + 2\,\Delta x, \cdots \qquad b,$$

define n equal intervals into which the whole interval $b-a$ is separated. Let $x_1, x_2, \cdots\cdots\cdots x_n$ be n values of x, one comprised in each of these intervals; also let $\Sigma_a^b f(x_r)\,\Delta x$ denote the sum of the n terms formed by giving to r the n values $1 \cdot 2 \cdots\cdots n$ in the typical term $f(x_r)\,\Delta x$; that is, let

$$\Sigma_a^b f(x_r)\,\Delta x = f(x_1)\,\Delta x + f(x_2)\,\Delta x \cdots\cdots + f(x_n)\,\Delta x. \quad . \quad . \quad (5)$$

We shall now show that when n is indefinitely increased the limiting value of $\Sigma_a^b f(x_r)\,\Delta x$ is $\int_a^b f(x)\,dx$.

99. If we separate the integral into parts corresponding to the terms above mentioned; thus,

$$\int_a^b f(x)\,dx = \int_a^{a+\Delta x} f(x)\,dx + \int_{a+\Delta x}^{a+2\Delta x} f(x)\,dx \cdots + \int_{b-\Delta x}^{b} f(x)\,dx,$$

and let $M_1, M_2, \cdots M_n$ denote the mean values of $f(x)$ in the several intervals, we have, in accordance with equation (3), Art. 97,

$$\int_a^b f(x)\,dx = M_1\Delta x + M_2\,\Delta x \cdots + M_n\,\Delta x. \quad . \quad . \quad . \quad . \quad (6)$$

Now, since $f(x_r)$ and M_r are both intermediate in value between the greatest and the least values of $f(x)$ in the interval to which they belong, their difference is less than the difference between these values of $f(x)$. Therefore, if we put

$$f(x_r) = M_r + e_r, \quad . \quad . \quad . \quad . \quad . \quad . \quad . \quad (7)$$

e_r is a quantity whose limit is zero when n, the number of intervals, is indefinitely increased, and Δx in consequence diminished indefinitely.

Comparing the terms in equations (5) and (6) we have, by means of equation (7),

$$\Sigma_a^b f(x)\,\Delta x = \int_a^b f(x)\,dx + (e_1 + e_2 \cdots + e_n)\,\Delta x. \quad . \quad . \quad . \quad (8)$$

Denote by e the arithmetical mean of the n quantities e_1, $e_2, \dots e_n$; that is, let

$$ne = e_1 + e_2 + e_3 \cdots e_n; \quad . \quad . \quad . \quad . \quad . \quad . \quad . \quad (9)$$

then, since e is an intermediate value between the greatest and the least value of e_r, it is also a quantity whose limit is zero when n is indefinitely increased. By equations (9) and (4), equation (8) becomes

$$\Sigma_a^b f(x_r) \Delta x = \int_a^b f(x)\, dx + e(b - a),$$

whence it follows that $\int_a^b f(x)\, dx$ is the limit of $\Sigma_a^b f(x_r)\, dx$ when n is indefinitely increased, since the limit of e is zero.

100. It was shown in the Differential Calculus, Art. 390 [Abridged Ed., Art. 193], that, in an expression for the ratio of finite differences, we may pass to the limit which the expression approaches, when the differences are diminished without limit, by substituting the symbol d for the symbol Δ. The theorem proved in the preceding articles shows that, in like manner, in the summation of an expression involving finite differences, we may pass to the limit approached when the differences are indefinitely diminished, by changing the symbols Σ and Δ into $\int$ and d.

The term *integral*, and the use of the long *s*, the initial of the word *sum*, as the sign of integration, have their origin in this connection between the processes of integration and summation.

Additional Formulas of Integration.

101. The formulas recapitulated below are useful in evaluating other integrals. (A) and (A') are demonstrated in Art. 17; (B) and (C) in Art. 29; (D) and (E) in Art. 30; (F) in Art. 31; (G) and (G') in Art. 35; (H) and (I) in Art. 50; (J) in Art. 51; (K) in Art. 52; (L) in Art. 53; (M) in Art. 55; (N) and (O) in Art. 58; (P) and (P') in Art. 84; and (Q) in Art. 85.

$$\int \frac{dx}{(x-a)(x-b)} = \frac{1}{a-b} \log \frac{x-a}{x-b} . \qquad (A)$$

$$\int \frac{dx}{x^2 - a^2} = \frac{1}{2a} \log \frac{x-a}{x+a} . \qquad (A')$$

$$\int \sin^2 \theta \, d\theta = \tfrac{1}{2} (\theta - \sin \theta \cos \theta) . \qquad (B)$$

$$\int \cos^2 \theta \, d\theta = \tfrac{1}{2} (\theta + \sin \theta \cos \theta) . \qquad (C)$$

$$\int \frac{d\theta}{\sin \theta \cos \theta} = \log \tan \theta . \qquad (D)$$

$$\int \frac{d\theta}{\sin \theta} = \log \tan \tfrac{1}{2}\theta = \log \frac{1 - \cos \theta}{\sin \theta} . \qquad (E)$$

$$\int \frac{d\theta}{\cos \theta} = \log \tan \left[\frac{\pi}{4} + \frac{\theta}{2} \right] = \log \frac{1 + \sin \theta}{\cos \theta} . \qquad (F)$$

$$\int \frac{d\theta}{a + b \cos \theta} = \frac{2}{\sqrt{(a^2 - b^2)}} \tan^{-1} \left[\sqrt{\frac{a-b}{a+b}} \tan \tfrac{1}{2}\theta \right] . \qquad (G)$$

$$\int \frac{d\theta}{a + b\cos\theta} = \frac{1}{\sqrt{(b^2 - a^2)}} \log \frac{\sqrt{(b+a)} + \sqrt{(b-a)}\tan\frac{1}{2}\theta}{\sqrt{(b+a)} - \sqrt{(b-a)}\tan\frac{1}{2}\theta} \quad . \; . \; (G')$$

$$\int \frac{dx}{x\sqrt{(x^2 + a^2)}} = \frac{1}{a} \log \frac{\sqrt{(x^2 + a^2)} - a}{x} \quad . \; . \; . \; . \; . \; . \; . \; . \; . \; (H)$$

$$\int \frac{dx}{x\sqrt{(a^2 - x^2)}} = \frac{1}{a} \log \frac{a - \sqrt{(a^2 - x^2)}}{x} \quad . \; . \; . \; . \; . \; . \; . \; . \; . \; (I)$$

$$\int \frac{dx}{(ax^2 + b)^{\frac{3}{2}}} = \frac{x}{b\sqrt{(ax^2 + b)}} \quad . \; . \; . \; . \; . \; . \; . \; . \; . \; . \; . \; . \; . \; (J)$$

$$\int \frac{dx}{\sqrt{(x^2 \pm a^2)}} = \log\left[x + \sqrt{(x^2 \pm a^2)}\right] \quad . \; . \; . \; . \; . \; . \; . \; . \; . \; . \; (K)$$

$$\int \sqrt{(x^2 \pm a^2)}\, dx = \frac{x\sqrt{(x^2 \pm a^2)}}{2} \pm \frac{a^2}{2} \log\left[x + \sqrt{(x^2 \pm a^2)}\right] \quad . \; . \; (L)$$

$$\int \sqrt{(a^2 - x^2)}\, dx = \frac{a^2}{2} \sin^{-1}\frac{x}{a} + \frac{x\sqrt{(a^2 - x^2)}}{2} \quad . \; . \; . \; . \; . \; . \; . \; (M)$$

$$\int \frac{dx}{\sqrt{[(x-\alpha)(x-\beta)]}} = 2\log\left[\sqrt{(x-\alpha)} + \sqrt{(x-\beta)}\right] \quad . \; . \; . \; (N)$$

$$\int \frac{dx}{\sqrt{[(x-\alpha)(\beta - x)]}} = 2\sin^{-1}\sqrt{\frac{x-\alpha}{\beta-\alpha}} \quad . \; . \; . \; . \; . \; . \; . \; . \; (O)$$

$$\int_0^{\frac{\pi}{2}} \sin^m\theta\, d\theta = \int_0^{\frac{\pi}{2}} \cos^m\theta\, d\theta = \frac{(m-1)(m-3)\cdots\cdots 1}{m(m-2)\cdots\cdots\cdots 2} \cdot \frac{\pi}{2} \quad . \; . \; (P)$$

$$\int_0^{\frac{\pi}{2}} \sin^m \theta \, d\theta = \int_0^{\frac{\pi}{2}} \cos^m \theta \, d\theta = \frac{(m-1)(m-3)\cdots\cdot 2}{m(m-2)\cdots\cdots 1} \quad . \quad . \quad . \quad (P')$$

$$\int_0^{\frac{\pi}{2}} \sin^m \theta \cos^n \theta \, d\theta = \frac{(m-1)(m-3)\cdots\times(n-1)(n-3)\cdots}{(m+n)(m+n-2)\cdots\cdots\cdots}\alpha, \quad . \ (Q)$$

in which $\alpha = 1$, unless m and n are both even, when $\alpha = \frac{\pi}{2}$.

Examples VII.

1. $\displaystyle\int_0^{n\pi} \frac{d\theta}{a + b\cos\theta}$, [$a > b$, and n an integer] $\qquad \dfrac{n\pi}{\sqrt{(a^2 - b^2)}}$.

2. $\displaystyle\int_0^{2n\pi \pm \frac{\pi}{2}} \frac{d\theta}{2 + \cos\theta}$, $\qquad \dfrac{2n\pi \pm \frac{1}{3}\pi}{\sqrt{3}}$.

3. $\displaystyle\int_0^{\frac{\pi}{2}} \sin^6 \theta \, d\theta$, $\qquad \dfrac{5\pi}{32}$.

4. $\displaystyle\int_0^{\pi} \sin^5 \theta \, d\theta$, $\qquad \dfrac{16}{15}$.

5. $\displaystyle\int_{-\frac{\pi}{2}}^{\pi} \cos^7 \theta \, d\theta$, $\qquad \dfrac{16}{35}$.

6. $\displaystyle\int_0^{\frac{\pi}{2}} \sin^4 \theta \cos^6 \theta \, d\theta$, $\qquad \dfrac{3\pi}{512}$.

7. $\int_0^{\pi} \sin^3 \theta \cos^4 \theta \, d\theta,$ $\qquad \frac{4}{35}.$

8. $\int_0^{\frac{\pi}{2}} \sin^m \theta \cos^m \theta \, d\theta,$ $\qquad \frac{1}{2^m} \int_0^{\frac{\pi}{2}} \sin^m \theta \, d\theta.$

9. $\int_0^1 \frac{x^{2n} \, dx}{\sqrt{(1-x^2)}},$ $\qquad \frac{1 \cdot 3 \cdot 5 \cdots (2n-1)}{2 \cdot 4 \cdot 6 \cdots 2n} \frac{\pi}{2}.$

10. $\int_0^1 \frac{x^{2n+1} \, dx}{\sqrt{(1-x^2)}},$ $\qquad \frac{2 \cdot 4 \cdot 6 \cdots 2n}{3 \cdot 5 \cdot 7 \cdots (2n+1)}.$

11. $\int_0^a x^3 (a^2 - x^2)^{\frac{5}{2}} \, dx,$ $\qquad \frac{2a^9}{63}.$

12. $\int_a^{\infty} \frac{(x^2 - a^2)^{\frac{3}{2}} \, dx}{x^6},$ $\qquad \frac{3\pi}{16a}.$

13. $\int_0^{\infty} \frac{x^5 \, dx}{(a^2 + x^2)^{\frac{7}{2}}},$ $\qquad \frac{8}{15a}.$

14. $\int_0^{\infty} \frac{x^4 \, dx}{(a^2 + x^2)^4},$ $\qquad \frac{\pi}{32a^3}.$

15. Prove that

$$\int_0^a x^{n-1} (a-x)^{m-1} \, dx = \int_0^a x^{m-1} (a-x)^{n-1} \, dx,$$

and derive a formula of reduction for this integral, supposing $n > 0$ and $m > 1$.

$$\int_0^a x^{n-1} (a-x)^{m-1} \, dx = \frac{m-1}{n} \int_0^a x^n (a-x)^{m-2} \, dx.$$

16. Deduce from the result of Ex. 15 the value of the integral when m is an integer.

$$\int_0^a x^{n-1}(a-x)^{m-1}\,dx = \frac{1\cdot2\cdot3\cdots(m-1)}{n(n+1)\cdots(n+m-1)}a^{m+n-1}.$$

17. $\int_{-a}^{a}(a+x)^6(a-x)^{\frac{3}{2}}\,dx.$ *See Ex.* 16. $\qquad \dfrac{2^{16}\sqrt{2}}{45045}a^{\frac{15}{2}}.$

18. $\int_0^{\frac{\pi}{2}}\sin^7\theta\,(\cos\theta)^{\frac{5}{2}}\,d\theta.$ *Put* $\sin^2\theta = x$, *and see Ex.* 16.

$$\frac{2^8}{5\cdot7\cdot11\cdot19}.$$

19. Show by a change of independent variable that

$$\int_0^\infty \frac{x^2\,dx}{(a^2+x^2)^2} = \int_0^\infty \frac{a^2\,dx}{(a^2+x^2)^2},$$

and therefore $\qquad \displaystyle\int_0^\infty \frac{x^2\,dx}{(a^2+x^2)^2} = \frac{1}{2}\int_0^\infty\frac{dx}{a^2+x^2} = \frac{\pi}{4a}.$

20. $\displaystyle\int_0^\infty \frac{x\log x\,dx}{(x^2+a^2)^2},$ $\qquad \dfrac{\log a}{2a^2}.$

21. $\displaystyle\int_0^\infty \frac{\tan^{-1}x\,.\,dx}{x^2+x+1},$ $\qquad \dfrac{\pi^2}{6\sqrt{3}}.$

22. $\displaystyle\int_0^\infty \tan^{-1}\frac{x}{a}\,\frac{x\,dx}{x^4+a^4},$ $\qquad \dfrac{\pi^2}{16a^2}.$

23. Derive a series of integrals by successive differentiation of the definite integral $\int_0^\infty \varepsilon^{-ax}\,dx.$

$$\int_0^\infty x^n\varepsilon^{-ax}\,dx = \frac{1\cdot2\cdots n}{a^{n+1}}.$$

24. Derive from the result of Art. 63 the definite integrals

$$\int_0^\infty \varepsilon^{-mx} \sin nx\, dx = \frac{n}{m^2 + n^2}, \quad \text{and} \quad \int_0^\infty \varepsilon^{-mx} \cos nx\, dx = \frac{m}{m^2 + n^2};$$

and thence derive by differentiation the integrals

$$\int_0^\infty x\varepsilon^{-mx} \sin nx\, dx = \frac{2mn}{(m^2 + n^2)^2}, \quad \text{and} \quad \int_0^\infty x\varepsilon^{-mx} \cos nx\, dx = \frac{m^2 - n^2}{(m^2 + n^2)^2}.$$

25. From the results of Ex. 24 derive

$$\int_0^\infty x^2 \varepsilon^{-mx} \sin nx\, dx = \frac{2n(3m^2 - n^2)}{(m^2 + n^2)^3};$$

$$\int_0^\infty x^2 \varepsilon^{-mx} \cos nx\, dx = \frac{2m(m^2 - 3n^2)}{(m^2 + n^2)^3}.$$

26. From the fundamental formula (k') derive

$$\int_0^\infty \frac{dx}{\alpha + \beta x^2} = \frac{\pi}{2\alpha^{\frac{1}{2}}\beta^{\frac{1}{2}}};$$

and thence derive a series of formulas by differentiation with reference to α.

$$\int_0^\infty \frac{dx}{(\alpha + \beta x^2)^n} = \frac{\pi}{2^n \beta^{\frac{1}{2}}} \cdot \frac{1 \cdot 3 \cdots (2n - 3)}{1 \cdot 2 \cdots (n - 1)} \cdot \frac{1}{\alpha^{n - \frac{1}{2}}}.$$

27. Derive a series of integrals by differentiating with reference to β, the integral used in Ex. 26.

$$\int_0^\infty \frac{x^{2n-2}\, dx}{(\alpha + \beta x^2)^n} = \frac{\pi}{2^n \alpha^{\frac{1}{2}}} \frac{1 \cdot 3 \cdot 5 \cdots (2n - 3)}{1 \cdot 2 \cdot 3 \cdots (n - 1)} \cdot \frac{1}{\beta^{n - \frac{1}{2}}}.$$

28. From the integral employed in examples 26 and 27, derive the value of $\int_0^\infty \frac{x^4\,dx}{(\alpha + \beta x^2)^4}$.

Differentiate twice with reference to β, and once with reference to α.

$$\int_0^\infty \frac{x^4\,dx}{(\alpha + \beta x^2)^4} = \frac{1\cdot 3\cdot 1}{1\cdot 2\cdot 3}\cdot\frac{\pi}{16\alpha^{\frac{3}{2}}\beta^{\frac{5}{2}}}.$$

29. Derive an integral by differentiation, from the result of Ex. II., 67.

$$\int_0^\infty \frac{dx}{(x^2 + b^2)(x^2 + a^2)^2} = \frac{\pi(2a + b)}{4a^3 b(a + b)^2}.$$

30. Derive an integral by integrating $\int_0^\infty \frac{dx}{a^2 + x^2} = \frac{\pi}{2a}$.

$$\int_0^\infty \left[\tan^{-1}\frac{p}{x} - \tan^{-1}\frac{q}{x}\right]\frac{dx}{x} = \frac{\pi}{2}\log\frac{p}{q}.$$

31. Derive a definite integral by integrating

$$\int_0^\infty \varepsilon^{-mx}\sin nx\,dx = \frac{n}{m^2 + n^2}$$

with reference to n.

$$\int_0^\infty \frac{\varepsilon^{-mx}}{x}(\cos ax - \cos bx)\,dx = \frac{1}{2}\log\frac{m^2 + b^2}{m^2 + a^2}.$$

32. Derive a definite integral from the integral employed in Ex. 31 by integration with reference to m.

$$\int_0^\infty \frac{\sin nx}{x}\left[\varepsilon^{-ax} - \varepsilon^{-bx}\right]dx = \left[\tan^{-1}\frac{b}{n} - \tan^{-1}\frac{a}{n}\right].$$

33. Derive an integral by integrating with respect to m

$$\int_0^\infty \varepsilon^{-mx} \cos nx\, dx = \frac{m}{m^2 + n^2}.$$

$$\int_0^\infty \frac{\varepsilon^{-ax} - \varepsilon^{-bx}}{x} \cos nx\, dx = \frac{1}{2} \log \frac{b^2 + n^2}{a^2 + n^2}.$$

34. Derive an integral by integrating with respect to n the integral used in the preceding example.

$$\int_0^\infty \frac{\varepsilon^{-mx}}{x} (\sin ax - \sin bx)\, dx = \tan^{-1} \frac{m(a - b)}{m^2 + ab}.$$

35. Show by means of the result of Ex. 32 that

$$\int_0^\infty \frac{\sin nx}{x} dx = \frac{\pi}{2}.$$

36. Derive an integral by integration from the result of Ex. II., 67.

$$\int_0^\infty \frac{1}{x} \left[\tan^{-1} \frac{p}{x} - \tan^{-1} \frac{q}{x}\right] \frac{dx}{x^2 + b^2} = \frac{\pi}{2b^2} \log \frac{p(q + b)}{q(p + b)}.$$

37. Evaluate $\int_0^\infty \log \frac{x^2 + a^2}{x^2 + b^2} dx$ by the method of Art. 96.

$$\pi (a - b).$$

38. Evaluate $\int_0^\infty \log \left[1 + \frac{a^2}{x^2}\right] \log x\, dx.$ $\qquad \pi a (\log a - 1).$

CHAPTER III.

GEOMETRICAL APPLICATIONS.

VIII.

Plane Areas.

102. THE first step in making an application of the Integral Calculus is to express the required magnitude in the form of an integral. In the geometrical applications, the magnitude is regarded as generated while some selected independent variable undergoes a given change of value. The independent variable is usually a straight line or an angle, varying between known limits; the required magnitude is either a line regarded as generated by the motion of a point, an area generated by the motion of a line, or a solid generated by the motion of an area. A plane area may be generated by the motion of a straight line, generally of variable length, the method selected depending upon the mode in which the boundaries of the area are defined.

An Area Generated by a Variable Line having a Fixed Direction.

103. The differential of the area generated by the ordinate of a curve, whose equation is given in rectangular coordinates, has been derived in Art. 3. The same method may be employed in the case of any area generated by a straight line whose direction is invariable.

Let AB be the generating line, and let R be its intersection with a fixed line CD, to which it is always perpendicular. Suppose R to move uniformly along CD, and let RS be the space described by R in the interval of time, dt. Then the value of the differential of the area, at the instant when the generating line passes the position AB, is the area which would be generated in the time dt, if the rate of the area were constant. This rate would evidently become constant if the generating line were made constant in length; and therefore the differential is the rectangle, represented in the figure, whose base and altitude are AB and RS; that is, it is *the product of the generating line, and the differential of its motion in a direction perpendicular to its length.*

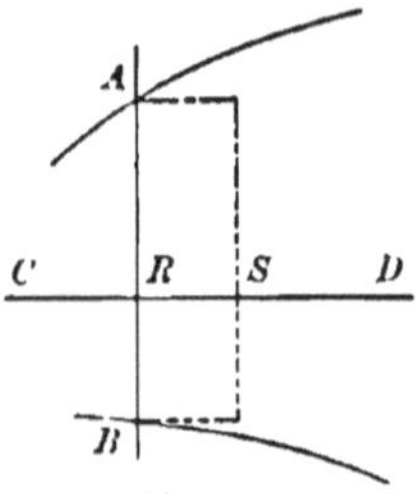

Fig. 3.

104. In the algebraic expression of this principle, the independent variable is the distance of R from some fixed origin upon CD, and the length of AB is to be expressed in terms of this independent variable.

When the curve or curves defining the length of AB are given in rectangular coordinates, CD is generally one of the axes; thus, if the generating line is the ordinate of a curve, the differential is $y\,dx$, as shown in Art. 3. It is often, however, convenient to regard the area as generated by some other line.

For example, given the curve known as the witch, whose equation is

$$y^2x - 2ay^2 + 4a^2x = 0 \quad . \quad . \quad . \quad . \quad . \quad . \quad . \quad (1)$$

This curve passes through the origin, is symmetrical to the axis of x, and has the line $x = 2a$ for an asymptote, since $x = 2a$ makes $y = \pm\infty$.

Let the area between the curve and its asymptote be required. We may regard this area as generated by the line PQ parallel to the axis of x, y being taken as the independent variable. Now

$$PQ = 2a - x,$$

hence the required area is

$$A = \int_{-\infty}^{\infty} (2a - x)\, dy \quad . \quad . \quad . \quad . \quad (2)$$

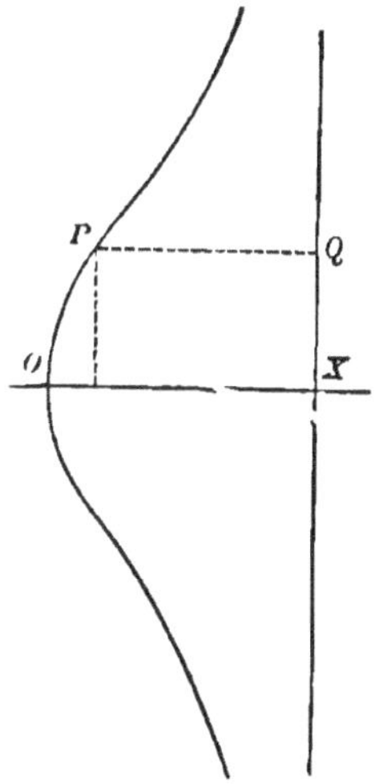

FIG. 4.

From the equation (1) of the curve, we have

$$x = \frac{2ay^2}{y^2 + 4a^2};$$

whence $\qquad 2a - x = \dfrac{8a^3}{y^2 + 4a^2},$

and equation (2) becomes

$$A = 8a^3 \int_{-\infty}^{\infty} \frac{dy}{y^2 + 4a^2} = 4a^2 \tan^{-1} \frac{y}{2a}\Big]_{-\infty}^{\infty} = 4\pi a^2.$$

Oblique Coordinates.

105. When the coordinate axes are oblique, if α denotes the angle between them, and the ordinate is the generating line, the differential of its motion in a direction perpendicular to its length is evidently $\sin \alpha \cdot dx$; therefore, the expression for the area is

$$A = \sin \alpha \int y\, dx.$$

As an illustration let the area between a parabola and a chord passing through the focus be required. It is shown in treatises on conic sections, the expression for a focal chord is

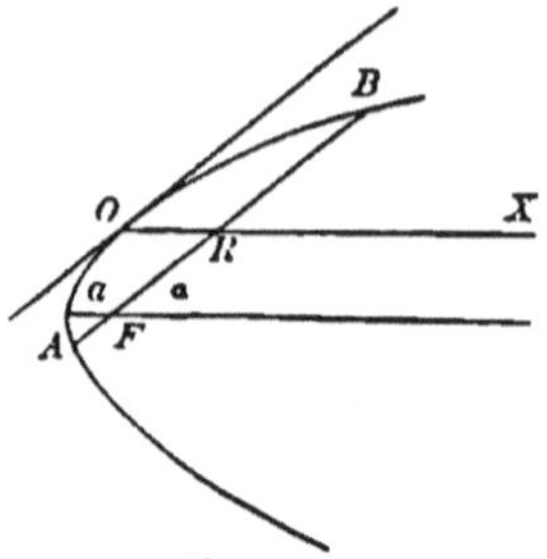

FIG. 5.

$$AB = 4a \operatorname{cosec}^2 \alpha, \quad . \quad . \quad . \quad (1)$$

where α is the inclination of the chord to the axis of the curve, and a is the distance from the focus to the vertex. It is also shown that the equation of the curve referred to the diameter which bisects the chord, and the tangent at its extremity which is parallel to the chord is

$$y^2 = 4a \operatorname{cosec}^2 \alpha \cdot x. \quad . \quad . \quad . \quad . \quad . \quad . \quad . \quad (2)$$

The required area may be generated by the double ordinate in this equation; and since from (1) the final value of y is $\pm\, 2a \operatorname{cosec}^2 \alpha$, equation (2) gives for the final value of x

$$OR = a \operatorname{cosec}^2 \alpha.$$

Hence we have

$$A = 2 \sin \alpha \int_0^{a \operatorname{cosec}^2 \alpha} y\, dx,$$

or by equation (2)

$$A = 4\sqrt{a} \int_0^{a \operatorname{cosec}^2 \alpha} \sqrt{x}\, dx = \frac{8a^2 \operatorname{cosec}^3 \alpha}{3}.$$

Employment of an Auxiliary Variable.

106. We have hitherto assumed that, in the expression

$$A = \int y\, dx,$$

x is taken as the independent variable, so that dx may be assumed constant; and it is usual to take the limits in such a manner that dx is positive. The resulting value of A will then have the sign of y, and will change sign if y changes sign.

It is frequently desirable, however, as in the illustration given below, to express both y and dx in terms of some other variable. When this is done, it is to be noticed that it is not necessary that dx should retain the same sign throughout the entire integral. The limits may often be so taken that the extremityof the generating ordinate must pass completely around a closed curve, and in that case it is easily seen that the complete integral, which represents the algebraic sum of the areas generated positively and negatively, will be the whole area of the closed curve.

107. As an illustration, let the whole area of the closed curve

$$\left(\frac{x}{a}\right)^{\frac{2}{3}} + \left(\frac{y}{b}\right)^{\frac{2}{3}} = 1,$$

represented in Fig. 6, be required. If in this equation we put

$$\left(\frac{x}{a}\right)^{\frac{1}{3}} = \sin\psi,$$

we shall have

$$\left(\frac{y}{b}\right)^{\frac{1}{3}} = \cos\psi;$$

whence $$x = a\sin^3\psi, \qquad \text{and} \qquad y = b\cos^3\psi. \quad . \quad . \quad (1)$$

Therefore $$\int y\,dx = 3ab\int \cos^4\psi \sin^2\psi\, d\psi.$$

Now if in this integral we use the limits 0 and 2π, the point determined by equation (1) describes the whole curve in the direction $ABCDA$. Hence we have for the whole area

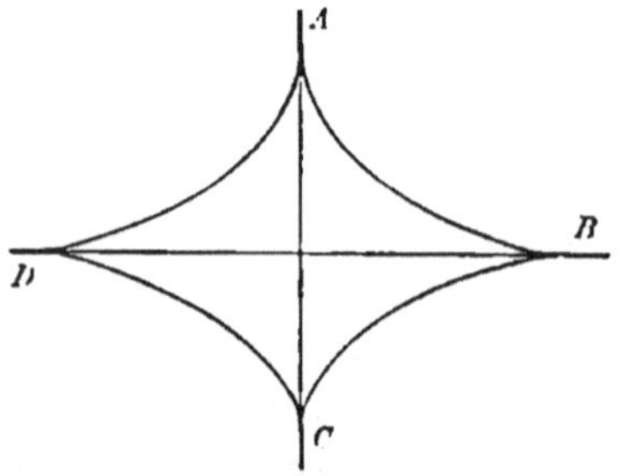

FIG. 6.

$$A = 3ab\int_0^{2\pi} \cos^4\psi \sin^2\psi\, d\psi,$$

and by formula (Q)

$$A = 3ab\,\frac{3\cdot 1\cdot 1}{6\cdot 4\cdot 2}\,2\pi = \frac{3\pi ab}{8}.$$

The areas in this case are all generated with the positive sign, since when y is negative dx is also negative. Had the generating point moved about the curve in the opposite direction, the result would have been negative.

Area generated by a Rotating Line or Radius Vector.

108. The radius vector of a curve given in polar coordinates is a variable line rotating about a fixed extremity. The angular rate is denoted by $\dfrac{d\theta}{dt}$ and may be regarded as constant, although the rate at which area is generated by the radius vector OP, Fig. 7, is not constant, because the length of OP is not constant. The differential of this area is the area which would be generated in the time dt, if the rate of the area were constant; that is to say, if the radius vector were of constant

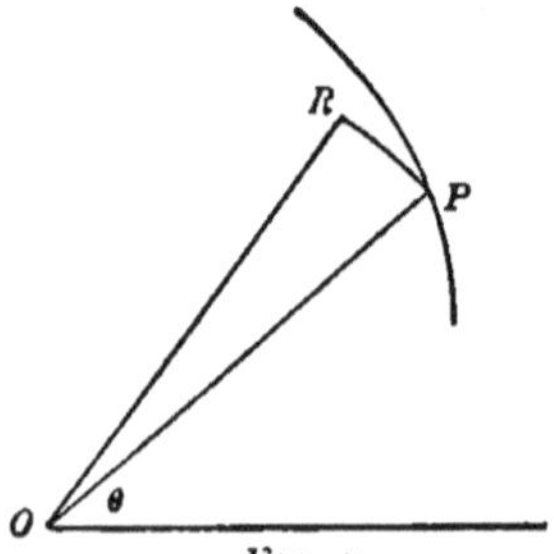

FIG. 7.

length. It is therefore the circular sector *OPR* of which the radius is r and the angle at the centre is $d\theta$.

Since
$$\text{arc } PR = r\, d\theta,$$
$$\text{sector } OPR = \frac{1}{2} r^2\, d\theta;$$

therefore the expression for the generated area is

$$A = \frac{1}{2}\int r^2\, d\theta \quad . \quad . \quad . \quad . \quad . \quad . \quad . \quad . \quad (1)$$

109. As an illustration, let us find the area of the right-hand loop of the lemniscata

$$r^2 = a^2 \cos 2\theta.$$

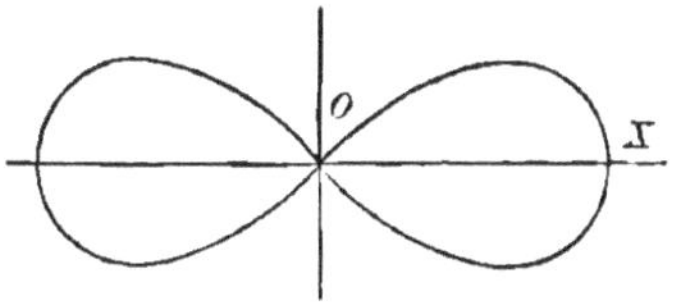

FIG. 8.

The limits to be employed are those values of θ which make $r = 0$; that is $-\frac{\pi}{4}$ and $\frac{\pi}{4}$.
Hence the area of the loop is

$$A = \frac{a^2}{2}\int_{-\frac{\pi}{4}}^{\frac{\pi}{4}} \cos 2\theta\, d\theta = \frac{a^2}{4} \sin 2\theta \Big]_{-\frac{\pi}{4}}^{\frac{\pi}{4}} = \frac{a^2}{2}.$$

110. When the radii vectores, r_2 and r_1 corresponding to the same value of θ in two curves, have the same sign, the area generated by their difference is the difference of the polar areas generated by r_1 and r_2. Hence the expression for this area is

$$A = \frac{1}{2}\int (r_2^2 - r_1^2)\, d\theta. \quad . \quad . \quad . \quad . \quad . \quad . \quad . \quad (2)$$

111. Let us apply this formula to find the whole area between the cissoid

$$r_1 = 2a\,(\sec\theta - \cos\theta),$$

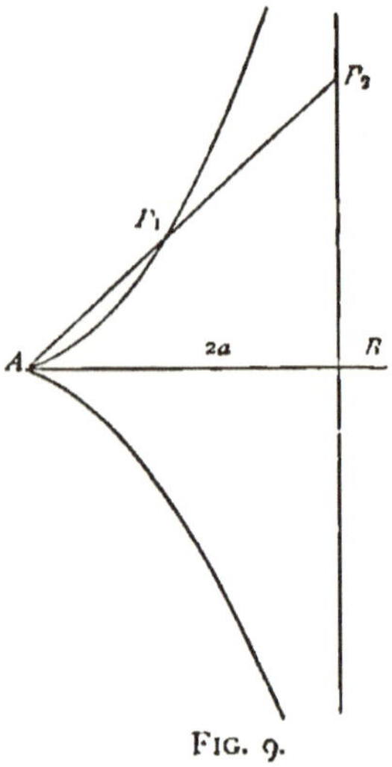

FIG. 9.

Fig. 9, and its asymptote BP_2, whose polar equation is

$$r_2 = 2a \sec\theta.$$

One half of the required area is generated by the line P_1P_2, while θ varies from 0 to $\frac{1}{2}\pi$. Hence by the formula

$$A = 2a^2\int_0^{\frac{\pi}{2}} (2-\cos^2\theta)\, d\theta = \frac{3}{2}\pi a^2.$$

Therefore the whole area required is $3\pi a^2$.

Transformation of the Polar Formulas.

112. In the case of curves given in rectangular coordinates, it is sometimes convenient to regard an area as generated by a radius vector, and to use the transformations deduced below in place of the polar formulas.

Put $$y = mx; \quad . \quad . \quad . \quad . \quad . \quad . \quad . \quad . \quad (1)$$

now taking the origin as pole and the initial line as the axis of x, we have

$$x = r\cos\theta, \qquad y = r\sin\theta; \quad . \quad . \quad . \quad (2)$$

therefore $$m = \frac{y}{x} = \tan\theta,$$

and $$dm = \sec^2\theta\, d\theta. \quad . \quad . \quad . \quad . \quad . \quad . \quad . \quad (3)$$

From equations (2) and (3),

$$x^2\,dm = r^2\,d\theta;$$

therefore equation (1) of Art. 108 gives

$$A = \frac{1}{2}\int x^2\,dm. \qquad (4)$$

In like manner, equation (2) Art. 110 becomes

$$A = \frac{1}{2}\int (x_2^2 - x_1^2)\,dm. \qquad (5)$$

113. As an illustration, let us take the folium

$$x^3 + y^3 - 3axy = 0. \qquad (1)$$

Putting $y = mx$, we have

$$x^3(1 + m^3) - 3amx^2 = 0. \qquad (2)$$

This equation gives three roots or values of x, of which two are always equal zero, and the third is

$$x = \frac{3am}{1 + m^3}; \qquad (4)$$

whence

$$y = \frac{3am^2}{1 + m^3}. \qquad (5)$$

These are therefore the coordinates of the point P in Fig. 10. Since the values of x and y vanish when $m = 0$, and when $m = \infty$, the curve has a loop in the first quadrant. To find the area of this loop we therefore have, by equation (4) of the preceding article,

$$A = \frac{9a^2}{2}\int_0^\infty \frac{m^2\,dm}{(1+m^3)^2} = -\frac{3a^2}{2}\frac{1}{1+m^3}\Big]_0^\infty = \frac{3a^2}{2}.$$

114. The area included between this curve and its asymptote may be found by means of equation (5), Art. 112. The equation of a straight line is of the form

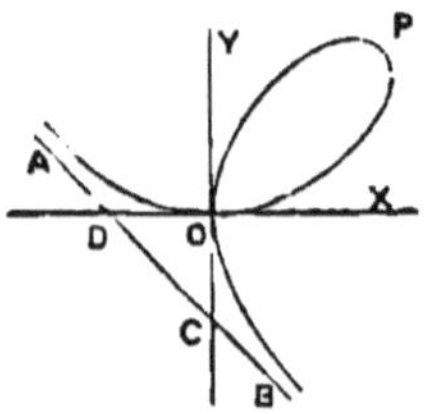

FIG. 10.

$$y = mx + b,$$

and since this line is parallel to $y = mx$, the value of m for the asymptote must be that which makes x and y in equations (4) and (5) infinite: that is, $m = -1$; hence the equation of the asymptote is

$$y + x = b, \quad . \quad . \quad . \quad . \quad . \qquad (6)$$

in which b is to be determined. Since when $m = -1$, the point P of the curve approaches indefinitely near to the asymptote, equation (6) must be satisfied by P when $m = -1$. From (4) and (5) we derive

$$y + x = 3a\frac{m^2+m}{1+m^3} = \frac{3am}{1-m+m^2};$$

whence, putting $m = -1$, and substituting in equation (6)

$$-a = b,$$

the equation of the asymptote AB, Fig. 10, is

$$y + x = -a. \quad . \quad . \quad . \quad . \quad . \quad . \quad . \quad . \quad . \qquad (7)$$

Now, as m varies from $-\infty$ to 0, the difference between the radii vectores of the asymptote and curve will generate the areas OBC and ODA, hence the sum of these areas is represented by

$$A = \frac{1}{2}\int_{-\infty}^{0} (x_2^2 - x_1^2)\, dm,$$

in which x_2 is taken from the equation of the asymptote (7), and x_1 from that of the curve.

Putting $y = mx$, in (7), we have

$$x_2 = -\frac{a}{1 + m},$$

and the value of x_1 is given in equation (4). Hence

$$A = \frac{a^2}{2}\int_{-\infty}^{0}\left[\frac{1}{(1 + m)^2} - \frac{9m^2}{(1 + m^3)^2}\right] dm$$

$$= \frac{a^2}{2}\left[\frac{3}{1 + m^3} - \frac{1}{1 + m}\right]_{-\infty}^{0}$$

$$= \frac{a^2}{2}\,\frac{2 + m - m^2}{1 + m^3}\Bigg]_{-\infty}^{0} = \frac{a^2}{2}\,\frac{2 - m}{1 - m + m^2}\Bigg]_{-\infty}^{0} {}^{*} = a^2.$$

Adding the triangle OCD, whose area is $\frac{1}{2}a^2$, we have for the whole area required $\frac{3}{2}a^2$.

---

* This reduction is given to show that the integral is not infinite for the value $m = -1$, which is between the limits. See Art. 77.

Examples VIII.

1. Find the area included between the curve

$$a^2y = x^3 + ax^2,$$

and the axis of x. $\frac{a^2}{12}$.

2. Find the whole area of the curve

$$a^4y^2 = x^4(a^2 - x^2). \qquad \frac{\pi a^2}{4}.$$

3. Find the area of a loop of the curve

$$x^2(a^2 + y^2) = y^2(a^2 - y^2). \qquad \frac{a^2}{2}(\pi - 2).$$

4. Find the area between the axes and the curve

$$y(x^2 + a^2) = b^2(a - x). \qquad b^2\left[\frac{\pi}{4} - \frac{\log 2}{2}\right].$$

5. Find the area between the curve

$$x^2y^2 + a^2y^2 - a^2x^2 = 0,$$

and one of its asymptotes. $2a^2$.

6. Find the area between the parabola $y^2 = 4ax$ and the straight line $y = x$. $\frac{8a^2}{3}$.

7. Find the area of the ellipse whose equation is

$$ax^2 + 2bxy + cy^2 = 1. \qquad \frac{\pi}{\sqrt{(ac - b^2)}}.$$

8. Find the area of the loop of the curve

$$cy^2 = (x - a)(x - b)^2,$$

in which $c > 0$ and $b > a$. $\dfrac{8(b - a)^{\frac{5}{2}}}{15\sqrt{c}}$.

9. Find the area of the loop of the curve

$$a^3y^2 = x^4(b + x). \qquad \frac{32b^{\frac{7}{2}}}{105a^{\frac{3}{2}}}:$$

10. Find the area included between the axes and the curve

$$\left(\frac{x}{a}\right)^{\frac{1}{3}} + \left(\frac{y}{b}\right)^{\frac{1}{3}} = 1. \qquad \frac{ab}{20}.$$

11. If n is an integer, prove that the area included between the axes and the curve

$$\left(\frac{x}{a}\right)^{\frac{1}{n}} + \left(\frac{y}{b}\right)^{\frac{1}{n}} = 1$$

is
$$A = \frac{n(n - 1)\cdots 1}{2n(2n - 1)\cdots(n + 1)}\,ab.$$

12. If n is an odd integer, prove that the area included between the axes and the curve

$$\left(\frac{x}{a}\right)^{\frac{2}{n}} + \left(\frac{y}{b}\right)^{\frac{2}{n}} = 1$$

is
$$A = \frac{[n(n - 2)\cdots 1]^2}{2n(2n - 2)\cdots 2}\,\frac{\pi ab}{2}.$$

13. In the case of the curtate cycloid

$$x = a\psi - b \sin \psi, \qquad y = a - b \cos \psi,$$

find the area between the axis of x and the arc below this axis.

$$(2a^2 + b^2) \cos^{-1}\frac{a}{b} - 3a\sqrt{(b^2 - a^2)}.$$

14. If $b = \frac{1}{2}a\pi$, show that the area of the loop of the curtate cycloid is

$$\pi a^2 \left[\frac{\pi^2}{8} - 1\right].$$

15. Find the area of the segment of the hyperbola

$$x = a \sec \psi, \qquad y = b \tan \psi,$$

cut off by the double ordinate whose length is $2b$.

$$ab\left[\sqrt{2} - \log \tan \frac{3\pi}{8}\right].$$

16. Find the whole area of the curve

$$r^2 = a^2 \cos^2 \theta + b^2 \sin^2 \theta. \qquad \frac{\pi}{2}(a^2 + b^2).$$

17. Find the area of a loop of the curve

$$r^2 = a^2 \cos^2 \theta - b^2 \sin^2 \theta. \qquad \frac{ab}{2} + \frac{(a^2 - b^2)}{2} \tan^{-1}\frac{a}{b}.$$

18. Find the areas of the large and of each of the small loops of the curve

$$r = a \cos \theta \cos 2\theta;$$

and show that the sum of the loops may be expressed by a single integral.

$$\frac{\pi a^2}{16} + \frac{a^2}{4}, \quad \text{and} \quad \frac{\pi a^2}{32} - \frac{a^2}{8}.$$

19. In the case of the spiral of Archimedes,

$$r = a\theta,$$

find the area generated by the radius vector of the first whorl and that generated by the difference between the radii vectores of the nth and $(n + 1)$th whorl.

$$\frac{8a^2\pi^3}{6}, \quad \text{and} \quad 8na^2\pi^3.$$

20. Find the area of a loop of the curve

$$r = a \sin 3\theta. \qquad \frac{\pi a^2}{12}.$$

21. Find the area of the cardioid

$$r = 4a \sin^2 \tfrac{1}{2}\theta. \qquad 6\pi a^2.$$

22. Find the area of the loop of the curve

$$r = a \frac{\cos 2\theta}{\cos \theta}. \qquad \frac{a^2(4 - \pi)}{2}.$$

23. In the case of the hyperbolic spiral,

$$r\theta = a,$$

show that the area generated by the radius vector is proportional to the difference between its initial and its final value.

24. Find the area of a loop of the curve

$$r = a \cos n\theta. \qquad \frac{\pi a^2}{4n}.$$

25. Find the area of a loop of the curve

$$r^2 = a^2 \frac{\sin 3\theta}{\cos^2\theta}. \qquad \frac{a^2}{2}.$$

26. Find the area of a loop of the curve

$$r^2 \sin\theta = a^2 \cos 2\theta.$$

Notice that r *is real and finite from* $\theta = \frac{5\pi}{4}$ *to* $\theta = \frac{7\pi}{4}$, *and that* $\int \frac{d\theta}{\sin\theta}$ *is negative in this interval.* $a^2\left[\sqrt{2} - \log(1 + \sqrt{2})\right].$

27. Find the area of a loop of the curve

$$(x^2 + y^2)^2 = a^2 xy.$$

Transform to polar coordinates. $\frac{a^2}{4}.$

28. In the case of the limaçon

$$r = 2a \cos\theta + b,$$

find the whole area of the curve when $b > 2a$ and show that the same expression gives the sum of the loops when $b < 2a$.

$(2a^2 + b^2)\pi.$

29. Find separately the areas of the large and small loops of the limaçon when $b < 2a$.

If $\alpha = \cos^{-1}\left(-\frac{b}{2a}\right)$,

$$\text{large loop} = (2a^2 + b^2)\,\alpha + \frac{3b}{2}\sqrt{(4a^2 - b^2)};$$

$$\text{small loop} = (2a^2 + b^2)\,(\pi - \alpha) - \frac{3b}{2}\sqrt{(4a^2 - b^2)}.$$

30. Find the area of a loop of the curve

$$r^2 = a^2 \cos n\theta + b^2 \sin n\theta. \qquad \frac{\sqrt{(a^4 + b^4)}}{n}.$$

31. Find the area of the loop of the curve

$$r = a\,\frac{2\cos 2\theta - 1}{\cos\theta}, \qquad \left[5\sqrt{3} - \frac{8}{3}\pi\right]a^2.$$

32. Show that the sectorial area between the axis of x, the equilateral hyperbola

$$x^2 - y^2 = 1,$$

and the radius vector making the angle θ at the centre is represented by the formula

$$A = \frac{1}{4}\log\frac{1 + \tan\theta}{1 - \tan\theta};$$

and hence show that

$$x = \frac{\varepsilon^{2A} + \varepsilon^{-2A}}{2}, \qquad \text{and} \qquad y = \frac{\varepsilon^{2A} - \varepsilon^{-2A}}{2}.$$

If A denotes the corresponding area in the case of the circle

$$x^2 + y^2 = 1,$$

we have

$$x = \cos 2A, \qquad \text{and} \qquad y = \sin 2A.$$

In accordance with the analogy thus presented, the values of x *and* y *given above are called the hyperbolic cosine and the hyperbolic sine of* 2A. *Thus*

$$\frac{\varepsilon^{2A} + \varepsilon^{-2A}}{2} = \cosh(2A), \qquad \frac{\varepsilon^{-2A} - \varepsilon^{2A}}{2} = \sinh(2A).$$

33. Find the area of the loop of the curve

$$x^4 - 3axy^2 + 2ay^3 = 0. \qquad \frac{3^6 a^2}{35 \cdot 2^4}.$$

34. Find the area of the loop of the curve

$$x^{2n+1} + y^{2n+1} = (2n+1)\, ax^n y^n. \qquad \frac{2n+1}{2} a^2.$$

35. Find the area between the curve

$$x^{2n+1} + y^{2n+1} = (2n+1)\, ax^n y^n$$

and its asymptote. $\frac{2n+1}{2} a^2.$

36. Find the area of the loop of the curve

$$y^3 + ax^2 - axy = 0. \qquad \frac{a^2}{60}$$

37. Find the area of a loop of the curve

$$x^4 + y^4 = a^2 xy. \qquad \frac{\pi a^2}{8}.$$

38. Trace the curve

$$x = 2a \sin \frac{y}{x},$$

and find the area of one loop. $\pi a^2.$

IX.

Volumes of Geometric Solids.

115. A geometric solid whose volume is required is frequently defined in such a way that the area of the plane section parallel to a fixed plane may be expressed in terms of the perpendicular distance of the section from the fixed plane. When this is the case, the solid is to be regarded as generated by the motion of the plane section, and its differential, when thus considered, is readily expressed.

116. For example, let us consider the solid whose surface is formed by the revolution of the curve APB, Fig. 11, about the axis OX. The plane section perpendicular to the axis OX is a circle; and if APB be referred to rectangular coordinates, the distance of the section from a parallel plane passing through the origin is x, while the radius of the circle is y. Supposing the centre of the section to move uniformly along the axis, the rate at which the volume is generated is not uniform, but its differential is the volume which would be generated while the centre is describing the distance dx, if the rate were made constant. This differential volume is therefore the cylinder whose altitude is dx, and the radius of whose base is y. Hence, if V denote the volume,

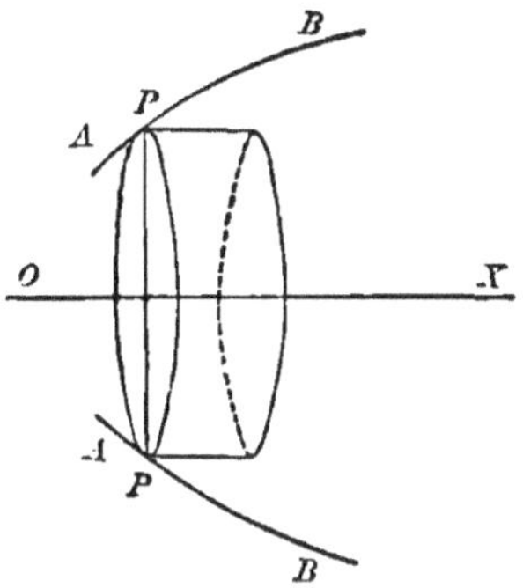

FIG. 11.

$$dV = \pi y^2\, dx.$$

117. As an illustration, let it be required to find the volume of the paraboloid, whose height is h, and the radius of whose base is b.

The revolving curve is in this case a parabola, whose equation is of the form

$$y^2 = 4ax;$$

and since $y = b$ when $x = h$,

$$b^2 = 4ah, \qquad \text{whence} \qquad 4a = \frac{b^2}{h};$$

the equation of the parabola is therefore

$$y^2 = \frac{b^2}{h}x.$$

Hence the volume required is

$$V = \pi \int_0^h y^2\,dx = \pi\,\frac{b^2}{h}\int_0^h x\,dx = \frac{\pi b^2 h}{2}.$$

118. It can obviously be shown, by the method used in Art. 116, that whatever be the shape of the section parallel to a fixed plane, *the differential of the volume is the product of the area of the generating section and the differential of its motion perpendicular to its plane.*

If the volume is completely enclosed by a surface whose equation is given in the rectangular coordinates x, y, z, and if we denote the areas of the sections perpendicular to the axes by A_x, A_y, and A_z, we may employ either of the formulas

$$V = \int A_x\,dx, \qquad V = \int A_y\,dy, \qquad V = \int A_z\,dz.$$

The equation of the section perpendicular to the axis of x is determined by regarding x as constant in the equation of the surface, and its area A_x is of course a function of x.

For example, the equation of the surface of an ellipsoid is

$$\frac{x^2}{a^2}+\frac{y^2}{b^2}+\frac{z^2}{c^2}=1.$$

The section perpendicular to the axis of x is the ellipse

$$\frac{y^2}{b^2}+\frac{z^2}{c^2}=\frac{a^2-x^2}{a^2},$$

whose semi-axes are $\frac{b}{a}\sqrt{(a^2-x^2)}$ and $\frac{c}{a}\sqrt{(a^2-x^2)}$.

Since the area of an ellipse is the product of π and its semi-axes,

$$A_x=\frac{\pi bc}{a^2}(a^2-x^2).$$

The limits for x are $\pm a$, the values between which x must lie to make the ellipse possible. Hence

$$V=\frac{\pi bc}{a^2}\int_{-a}^{a}(a^2-x^2)\,dx=\frac{4\pi abc}{3}.$$

119. The area A_x can frequently be determined by the conditions of the problem without finding the equation of the surface. For example, let it be required to find the volume of the solid generated by so moving an ellipse with constant major axis, that its center shall describe the major axis of a fixed ellipse, to whose plane it is perpendicular, while the extremities of its minor axis describe the fixed ellipse. Let the equation of the fixed ellipse be

$$\frac{x^2}{a^2}+\frac{y^2}{b^2}=1,$$

and let c be the major semi-axis of the moving ellipse. The minor semi-axis of this ellipse is y. Since the area of an ellipse is equal to π multiplied by the product of its semi-axes, we have

$$A_x = \pi c y = \frac{\pi c b}{a} \sqrt{(a^2 - x^2)}.$$

Therefore $$V = \frac{\pi b c}{a} \int_{-a}^{a} \sqrt{(a^2 - x^2)}\, dx;$$

hence, see formula (M),

$$V = \frac{\pi^2 abc}{2}.$$

The Solid of Revolution regarded as Generated by a Cylindrical Surface.

120. A solid of revolution may be generated in another manner, which is sometimes more convenient than the employment of a circular section, as in Art. 116. For example, let the cissoid POR, Fig. 12, whose equation is

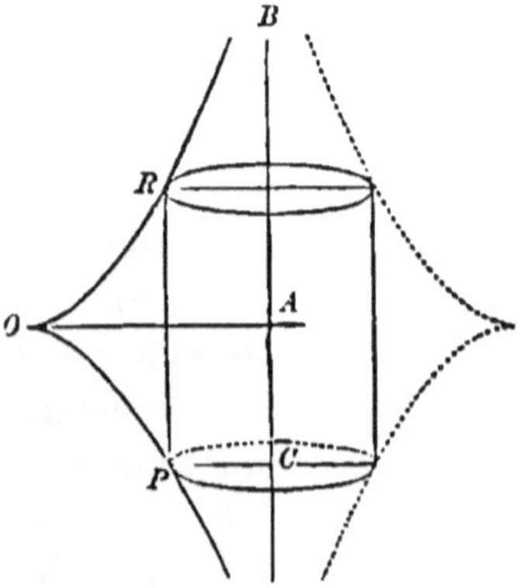

FIG. 12.

$$y^2 (2a - x) = x^3,$$

revolve about its asymptote AB. The line PR, parallel to AB and terminated by the curve, describes a cylindrical surface. If we conceive the radius of this cylinder to pass from the value $OA = 2a$ to zero, the cylindrical surface will evidently generate the solid of revolution. Now every point of this cylindrical surface moves with a rate equal to that of the radius; therefore the differential of the solid is the product of the cylindrical surface, and the differential of the radius. The radius and altitude in this case are

$$PC = 2a - x, \qquad \text{and} \qquad PR = 2y,$$

therefore
$$V = 4\pi \int_0^{2a} (2ax - x^2)^{\frac{1}{2}} x \, dx.$$

Putting
$$x - a = a \sin \theta,$$

$$V = 4\pi a^3 \int_{-\frac{\pi}{2}}^{\frac{\pi}{2}} (\cos^2 \theta + \cos^2 \theta \sin \theta) \, d\theta = 2\pi^2 a^3.$$

Double Integration.

121. When rectangular coordinates are used, the expression for the area generated by a line parallel to the axis of y and terminated by two curves is

$$A = \int_a^b (y_2 - y_1) \, dx. \quad . \quad . \quad . \quad (1)$$

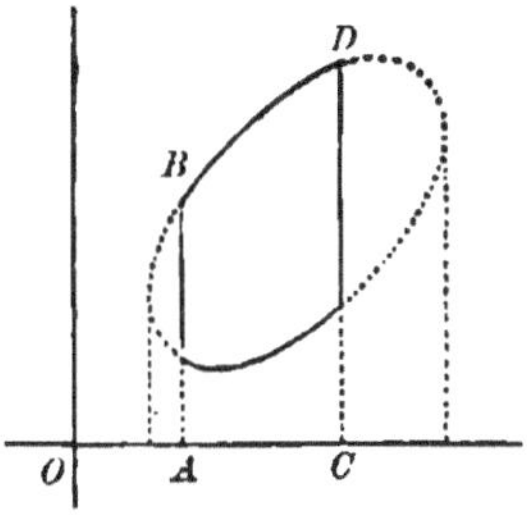

FIG. 13.

Let AB, in Fig. 13, be the initial, and CD the final position of the generating line, then the area is $ABDC$, which is enclosed by the curves

$$y = y_1, \qquad y = y_2,$$

and by the straight lines

$$x = a, \qquad x = b.$$

If in equation (1) we substitute for $y_2 - y_1$ the equivalent expression $\int_{y_1}^{y_2} dy$, we have

$$A = \int_a^b \int_{y_1}^{y_2} dy\, dx, \quad . \quad . \quad . \quad . \quad . \quad . \quad . \quad (2)$$

which expresses the area in the form of a double integral. In this double integral the limits y_1 and y_2 for y, are functions of x, while a and b, the limits for x, are constants.

122. If the area is that of a closed curve y_1 and y_2 are two values of y corresponding to the same value of x in the equation of the curve, and a and b are the values of x for which y_1 and y_2 become equal, as represented by the dotted lines in Fig. 13. It is evident that the entire area may also be expressed in the form

$$A = \int_p^q \int_{x_1}^{x_2} dx\, dy; \quad . \quad . \quad . \quad . \quad . \quad . \quad . \quad . \quad (3)$$

and that when either of the forms (2) or (3) is applied to the area of a closed curve the limits are completely determined by the equation of the curve.

123. The limits in either of the expressions (1) or (2) define a certain closed boundary, and since either of these integrals represents the included area, it is evident that we may write

$$\iint dy\, dx = \iint dx\, dy;$$

provided it is understood that the limits in the two expressions are such as to represent the same boundary. It should however be noticed that if the boundary is like that represented by the full lines in Fig. 13, or if the arcs $y = y_1$ and $y = y_2$ *do not belong to the same curve*, we cannot make a practical application of the form (3) without breaking up the integral into several parts.

124. Let $\phi(x, y)$ be any function of x and y. In the double integral

$$\int_a^b \int_{y_2}^{y_1} \phi(x, y)\, dy\, dx, \quad . \quad . \quad . \quad . \quad . \quad . \quad (1)$$

x is considered as a constant or independent of y in the first integration, but the limits of this integration are functions of x. The double integration is then said to *extend over the area* which is represented by the expression

$$\int_a^b \int_{y_1}^{y_2} dy\, dx, \qquad \text{or} \qquad \int_a^b (y_2 - y_1)\, dx. \quad . \quad . \quad . \quad . \quad (2)$$

125. Now let the surface, of which

$$z = \phi(x, y) \quad . \quad . \quad . \quad . \quad . \quad . \quad . \quad . \quad . \quad (3)$$

is the equation in rectangular coordinates, be constructed; and let a cylindrical surface be formed by moving a line perpendicular to the plane of xy about the boundary of the area (2) over which the integration extends. Let us suppose the value of z to be positive for all values of x and y which represent points within this boundary. Then the cylindrical surface, together with the plane of xy and the surface (3), encloses a solid, of which the base is the area (2) in the plane xy, or $ASBR$ in Fig. 14, and the upper surface is $CQDP$ a portion of the surface (3).

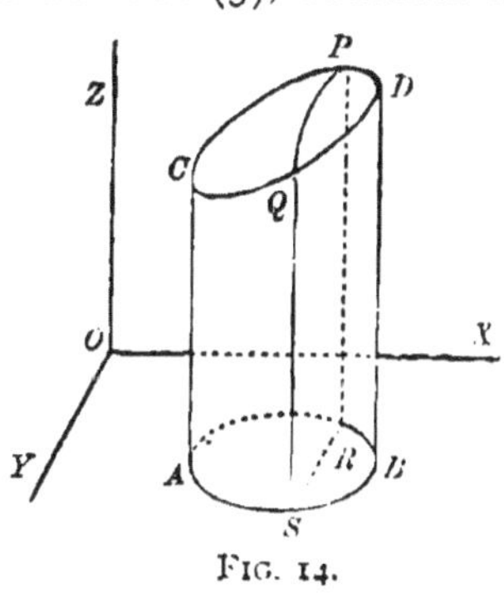

FIG. 14.

Let $SRPQ$ be a section of this solid perpendicular to the axis of x. In this section x has a constant value, and the ordinates of R and S are the corresponding values of y_1 and y_2. The area of this section, which denote by A_x, as in Art. 117, may be regarded as generated by the line z, hence

$$A_x = \int_{y_1}^{y_2} z\, dy;$$

and therefore

$$V = \int_a^b \int_{y_1}^{y_2} z\, dy\, dx, \quad . \quad . \quad . \quad . \quad . \quad . \quad . \quad . \quad (1)$$

which is identical with expression (1) Art. 124.

126. Now it is evident that the same volume may be expressed by

$$V = \iint z\, dx\, dy,$$

provided that the double integration extends over the same area. Hence, with this understanding, we may write

$$\iint \phi(x, y)\, dy\, dx = \iint \phi(x, y)\, dx\, dy.$$

In this formula x and y may be regarded as taking the places of any two variables, the limits of integration being determined by a given relation between the variables. Thus we may write

$$\iint \phi(u, v)\, dv\, du = \iint \phi(u, v)\, du\, dv,$$

provided the limits of integration are determined in each case by the same relation between u and v.

127. For example, if this relation is

$$u^2 + v^2 - c^2 = 0,$$

the range of values in the first integration is between

$$v = \pm \sqrt{(c^2 - u^2)};$$

that is, we must have

$$v^2 < c^2 - u^2,$$

or $$u^2 + v^2 - c^2 < 0 \quad . \quad . \quad . \quad . \quad . \quad . \quad . \quad . \quad (1)$$

But this condition also expresses the limits for u, since v is only possible when $u^2 < c^2$. Now, putting rectangular coordinates, x and y, in place of u and v, it is convenient to express the restriction (1), by saying that the range of values of x and y is such as to represent every point *within* the circle

$$x^2 + y^2 - c^2 = 0.$$

Volumes by Double and Triple Integration.

128. As an application of formula (1), Art. 125, let us suppose the curve $ASBR$ to be the circle

$$(x - h)^2 + (y - k)^2 = c^2, \quad . \quad . \quad . \quad . \quad . \quad . \quad . \quad (1)$$

and the equation of the surface $CQDP$ to be

$$xy = pz. \quad . \quad . \quad . \quad . \quad . \quad . \quad . \quad . \quad . \quad . \quad (2)$$

Then $$V = \frac{1}{p}\int_a^b \int_{y_1}^{y_2} xy\,dy\,dx = \frac{1}{2p}\int_a^b (y_2^2 - y_1^2)\,x\,dx,$$

in which the limits y'_1 and y'_2 are derived from equation (1). Hence

$$y'_2 = k + \sqrt{[c^2 - (x-h)^2]}, \qquad y'_1 = k - \sqrt{[c^2 - (x-h)^2]},$$

and

$$V = \frac{2k}{p}\int_a^b \sqrt{[c^2 - (x-h)^2]}\, x\, dx.$$

The limits for x are the extreme values of x which make y possible; that is,

$$a = h - c \qquad \text{and} \qquad b = h + c.$$

To evaluate the integral, put

$$x - h = c \sin\theta;$$

then

$$V = \frac{2kc^2}{p}\int_{-\frac{\pi}{2}}^{\frac{\pi}{2}} \cos^2\theta\,(h + c\sin\theta)\, d\theta.$$

Since, by Art. 87,

$$\int_{-\frac{\pi}{2}}^{\frac{\pi}{2}} \cos^2\theta \sin\theta\, d\theta = 0,$$

we have finally

$$V = \frac{\pi khc^2}{p}.$$

129. A volume in general may be represented by the triple integral

$$V = \iiint dz\, dy\, dx, \quad \ldots \ldots \ldots \quad (1)$$

which is equivalent to

$$V = \iint (z_2 - z_1)\, dy\, dx\,; \quad . \quad . \quad . \quad . \quad . \quad . \quad (2)$$

for $\int (z_2 - z_1)\, dy = A_x$, the area of a section perpendicular to the axis of x. We may regard this formula as expressing the difference between two cylindrical solids of the form represented in Fig. 14.

130. When the volume is that of a closed surface, z_2 and z_1 are two values of z in terms of x and y found from the equation of the surface. The area over which the integration extends is in this case the projection of the solid upon the plane of xy; in other words, the base of a circumscribing cylinder. Thus, if the volume is that of the sphere

$$x^2 + y^2 + (z - c)^2 = a^2, \quad . \quad . \quad . \quad . \quad . \quad . \quad (1)$$

z_1 and z_2 are the two values of z derived from this equation;

that is $$c \pm \sqrt{(a^2 - x^2 - y^2)}.$$

Hence $$z_2 - z_1 = 2\sqrt{(a^2 - x^2 - y^2)},$$

and $$V = 2\iint \sqrt{(a^2 - x^2 - y^2)}\, dy\, dx. \quad . \quad . \quad . \quad . \quad (2)$$

The integration here extends over the circle

$$x^2 + y^2 - a^2 = 0. \quad . \quad . \quad . \quad . \quad . \quad . \quad . \quad . \quad (3)$$

since $z_2 - z_1$ is real only when

$$a^2 - x^2 - y^2 > 0.$$

From equation (3) we find the limits for y to be

$$\pm \sqrt{(a^2 - x^2)},$$

hence, by formula (M), equation (2) becomes

$$V = \pi \int (a^2 - x^2)\, dx.$$

Finally the limits for x are $\pm a$, since y is real only when x is between these limits;

therefore $$V = \pi \left[a^2 x - \frac{1}{3} x^3 \right]_{-a}^{a} = \frac{4}{3} \pi a^3.$$

Elements of Area and Volume.

131. In accordance with Art. 100, the expression for an area,

$$\int_a^b \int_{y_1}^{y_2} dy\, dx, \quad . \quad . \quad . \quad . \quad . \quad . \quad . \quad . \quad (1)$$

is the limit of the sum

$$\Sigma_a^b [\Sigma_{y_1}^{y_2} \Delta y]\, \Delta x.$$

Since each of the terms included in $\Sigma_{y_1}^{y_2} \Delta y$ is multiplied by the common factor Δx, this sum may be written in the form

$$\Sigma_a^b \Sigma_{y_1}^{y_2} \Delta y\, \Delta x. \quad . \quad . \quad . \quad . \quad . \quad . \quad . \quad . \quad (2)$$

The sum (2) consists of terms of the form

$$\Delta y' \Delta x;$$

and this product is called *the element of the sum;* in like manner, the product

$$dy\, dx,$$

which takes the place of $\Delta y\, \Delta x$ when we pass to the limit by substituting integration for summation, is called *the element of the integral* (1), or of the area represented by it.

132. We may now regard the process of double integration as a process of double summation, as indicated by expression (2), followed by the act of passing to the limiting value. In the first summation indicated, the elemental rectangles corresponding to the same value of x are combined into the term $(y'_2 - y'_1)\,\Delta x$, which may be called a *linear element of area,* since its length is independent of the symbol Δ.

133. It is easy to see that, in a similar manner, when rectangular coordinates are used, a volume may be regarded as the limiting value of the sum of terms of the form

$$\Delta x\, \Delta y\, \Delta z;$$

and hence $\qquad dx\, dy\, dz,$

which takes its place when we pass to the limiting value by substituting integration for summation, is called *the element of volume.*

If the summation is effected in the order z, y, x, the first operation combines the elements which have common values of y and x into the *linear element of volume,*

$$(z_2 - z_1)\,\Delta x\, \Delta y.$$

The second operation combines the linear elements corresponding to a common value of x, over a certain range of values of y, into a term whose limiting value takes the form

$$A_x \Delta x.$$

This last expression represents a *lamina* perpendicular to the axis of x, whose area is A_x a section of the solid, and whose thickness is Δx.

Polar Elements.

134. If in the formula for a polar area,

$$A = \frac{1}{2}\int (r_2^2 - r_1^2)\, d\theta, \quad . \quad . \quad . \quad . \quad . \quad . \quad . \quad (1)$$

[equation (2), Art. 110], we substitute for $\frac{1}{2}(r_2^2 - r_1^2)$ the equivalent expression $\int_{r_1}^{r_2} r\, dr$, we obtain

$$A = \int_\alpha^\beta \int_{r_1}^{r_2} r\, dr\, d\theta, \quad . \quad . \quad . \quad . \quad . \quad . \quad (2)$$

in which α and β are fixed limits for θ.

Now it follows, from Art. 126, that the limits being determined by a certain relation between r and θ, this integral may also be put in the form

$$A = \int_a^b r \int_{\theta_1}^{\theta_2} d\theta \cdot dr = \int_a^b r\,(\theta_2 - \theta_1)\, dr, \quad . \quad . \quad (3)$$

in which a and b are the limiting values of r, between which θ is possible.

The expression $$r\,dr\,d\theta,$$

in equation (2), is called the *polar element of area.**

135. The formula

$$A = \int r\,(\theta_2 - \theta_1)\,dr$$

may also be derived geometrically; for $r\,(\theta_2 - \theta_1)$ is the length of an arc whose radius is r. As r increases, this arc generates the surface, and it is plain that every point has a motion, whose differential is dr, in a direction perpendicular to the arc.

136. In determining the volume of a solid, it is sometimes convenient to express z as a function of the polar coordinates of its projection in the plane of xy. In this case we employ the linear element of volume,

$$(z_2 - z_1)\,r\,dr\,d\theta,$$

corresponding to the polar element of area.

---

* It is easily shown that the area included between the circles whose radii are r and $r + \Delta r$, and the radii whose inclinations to the initial line are θ and $\theta + \Delta\theta$ is

$$(r + \tfrac{1}{2}\Delta r)\,\Delta r\,\Delta\theta.$$

Since $r + \frac{1}{2}\Delta r$ is intermediate between r and $r + \Delta r$, the limiting value of the sum, of which this is the element, is, by Art. 99, the integral of the element

$$r\,dr\,d\theta.$$

In the summation corresponding to equation (1), the elements are first combined into the *sectorial element*

$$\tfrac{1}{2}(r_2^2 - r_1^2)\,\Delta\theta;$$

while in the summation corresponding to equation (3), they are first combined into the *arc-shaped element*

$$(r + \tfrac{1}{2}\Delta r)(\theta_2 - \theta_1)\,\Delta r.$$

As an illustration, let us determine the volume cut from a sphere by a right cylinder, having a radius of the sphere for one of its diameters. Taking the centre of the sphere as the origin, the diameter of the cylinder as initial line, and the axis of z parallel to the axis of the cylinder, we have for every point on the surface of the sphere

$$z^2 + r^2 = a^2, \quad \ldots \ldots \quad (1)$$

where a is the radius of the sphere. Hence

$$z_2 - z_1 = 2\sqrt{(a^2 - r^2)},$$

and $$V = 2\iint_{r_1}^{r_2} (a^2 - r^2)^{\frac{1}{2}}\, r\, dr\, d\theta = \int \left[-\frac{2}{3}(a^2 - r^2)^{\frac{3}{2}} \right]_{r_1}^{r_2} d\theta.$$

The circular base passes through the pole, and its equation is

$$r = a \cos\theta, \quad \ldots \ldots \quad (2)$$

hence the limits for r are 0 and $a \cos\theta$, and by substitution we obtain

$$V = \frac{2a^3}{3}\int (1 - \sin^3\theta)\, d\theta.$$

The limits for θ are $\pm\frac{\pi}{2}$, the values which make r vanish in equation (2); but it is to be noticed that the expression $(a^2 - r^2)^{\frac{3}{2}}$, for which we have substituted $a^3 \sin^3\theta$, is *always positive*, whereas $\sin^3\theta$ is negative in the fourth quadrant. Hence the value of V is double the value of the integral in the first quadrant; that is,

$$V = \frac{4a^3}{3}\int_0^{\frac{\pi}{2}} (1 - \sin^3\theta)\, d\theta = \frac{2\pi a^3}{3} - \frac{8a^3}{9}.$$

If a second cylinder whose diameter is the opposite radius of the sphere be constructed, the whole volume removed from the sphere is $\dfrac{4\pi a^3}{3} - \dfrac{16a^3}{9}$, and the portion of the sphere which remains is $\dfrac{16a^3}{9}$, a quantity commensurable with the cube of the diameter.

Polar Coordinates in Space.

137. A point in space may be determined by the polar coordinates ρ, ϕ, and θ, of which ρ denotes the radius vector OP, Fig. 15, ϕ the inclination POR of ρ to a fixed plane passing through the pole, and θ the angle ROA, which the projection of ρ upon this plane makes with a fixed line in the plane. The angles ϕ and θ thus correspond to the latitude and longitude of the point P considered as situated upon the surface of a sphere whose radius is ρ. The radius of the circle of latitude BP is

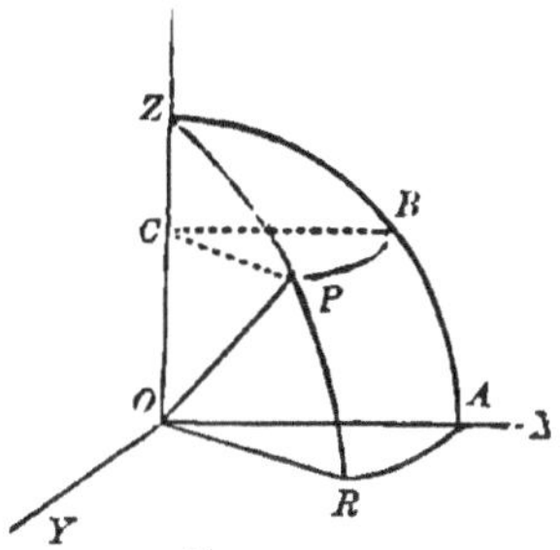

FIG. 15.

$$PC = \rho \cos \phi.$$

The motions of P, when ρ, ϕ, and θ independently vary, are in the directions of the radius vector OP and of the tangents at P to the arcs PR and PB. The differentials of these motions are respectively

$$d\rho, \qquad \rho\, d\phi, \qquad \text{and} \qquad \rho \cos \phi\, d\theta;$$

and since these motions are mutually rectangular, the element of volume is their product,

$$\rho^2 \cos\phi \, d\rho \, d\phi \, d\theta,$$

$$\text{and } V = \iiint \rho^2 \cos\phi \, d\rho \, d\phi \, d\theta. \quad . \quad . \quad . \quad . \quad . \quad (1)$$

138. Performing the integration with respect to ρ, the formula becomes

$$V = \frac{1}{3}\iint (\rho_2^3 - \rho_1^3) \cos\phi \, d\phi \, d\theta. \quad . \quad . \quad . \quad . \quad . \quad (2)$$

When the radius vector lies entirely within the solid, the lower limit ρ_1 must be taken equal zero, and we may write

$$V = \frac{1}{3}\iint \rho^3 \cos\phi \, d\phi \, d\theta. \quad . \quad . \quad . \quad . \quad . \quad . \quad . \quad (3)$$

The element of this double integral has the form of a pyramid with vertex at the pole.

If, on the other hand, in formula (1) we perform first the integration with respect to ϕ, we have

$$V = \iint (\sin\phi_2 - \sin\phi_1) \, \rho^2 \, d\rho \, d\theta. \quad . \quad . \quad . \quad . \quad . \quad (4)$$

Taking the lower limit $\phi_1 = 0$, so that the solid is bounded by the plane ORA, we have the simpler formula

$$V = \iint \sin\phi \, \rho^2 \, d\rho \, d\theta. \quad . \quad . \quad . \quad . \quad . \quad . \quad . \quad (5)$$

139. The formulas of the preceding article take simpler

forms when applied to solids of revolution. Let OZ, Fig. 15, be the axis of revolution, then ρ and ϕ are polar coordinates of the revolving curve, OR being the initial line. Now θ is in this case independent of ρ and ϕ, and its limits are 0 and 2π. The integration with reference to θ may therefore be performed at once. Thus from (3) we obtain

$$V = \frac{2\pi}{3}\int \rho^3 \cos\phi \, d\phi \, ; \quad . \quad . \quad . \quad . \quad . \quad . \quad (6)$$

and in each of the formulas the factor 2π may take the place of the integration with reference to θ.

140. As an example of the use of equation (6), let us find the volume generated by a circle revolving about one of its tangents. The initial line, being perpendicular to the axis of revolution, is a diameter; hence if a is the radius of the circle its equation is

$$\rho = 2a \cos\phi,$$

and the limits for ϕ are $-\frac{\pi}{2}$ and $\frac{\pi}{2}$. Substituting in (6)

$$V = \frac{16\pi a^3}{3}\int_{-\frac{\pi}{2}}^{\frac{\pi}{2}} \cos^4\phi \, d\phi = 2\pi^2 a^3.$$

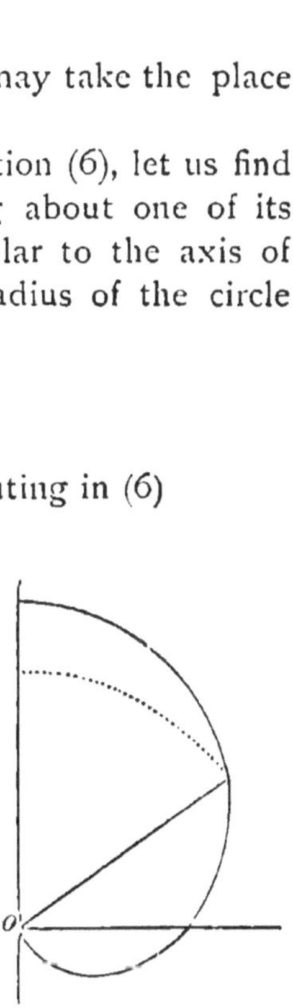

FIG. 16.

141. The following example of the use of equation (4), Art. 138, is added to illustrate the necessity of drawing a figure in each case to determine the limits to be employed.

Let it be required to find the volume generated by the revolution of the cardioid about its axis, the equation of the curve being

$$\rho = a(1 + \sin\phi), \quad . \quad . \quad . \quad . \quad . \quad . \quad . \quad (1)$$

when the initial line is perpendicular to the axis of the curve, as in Fig. 16. The figure shows that the upper limit for ϕ is $\frac{1}{2}\pi$, while the lower limit is the value of ϕ given by equation (1); therefore

$$\sin\phi_2 = 1, \qquad \text{and} \qquad \sin\phi_1 = \frac{\rho}{a} - 1.$$

The limits for ρ are evidently 0 and $2a$. Substituting in equation (4) Art. 138,

$$V = 2\pi \int_0^{2a} \left(2 - \frac{\rho}{a}\right) \rho^2 \, d\rho$$

$$= 2\pi \left[\frac{2\rho^3}{3} - \frac{\rho^4}{4a}\right]_0^{2a} = \frac{8\pi a^3}{3}.$$

Examples IX.

1. Find the volume of the spheroid produced by the revolution of the ellipse,

$$\frac{x^2}{a^2} + \frac{y^2}{b^2} = 1,$$

about the axis of x. $\frac{4\pi ab^2}{3}$.

2. Find the volume of a right cone whose altitude is a, and the radius of whose base is b. $\frac{\pi ab^2}{3}$.

3. Find the volume of the solid produced by the revolution about the axis of x of the area between this axis, the cissoid

$$y^2(2a - x) = x^3,$$

and the ordinate of the point (a, a). $8a^3\pi(\log 2 - \frac{2}{3})$.

4. Find the volume generated by the revolution of the witch,

$$y^2x - 2ay^2 + 4a^2x = 0,$$

about its asymptote.

See Art. 104. $4\pi^2a^3$.

5. The equilateral hyperbola

$$x^2 - y^2 = a^2$$

revolves about the axis of x: show that the volume cut off by a plane cutting the axis of x perpendicularly at a distance a from the vertex is equal to a sphere whose radius is a.

6. An anchor ring is formed by the revolution of a circle whose radius is b about a straight line in its plane at a distance a from its centre: find its volume. $2\pi^2ab^2$.

7. Express the volume of a segment of a sphere in terms of the altitude h and the radii a_1 and a_2 of the bases.

$$\frac{\pi h}{6}(h^2 + 3a_1^2 + 3a_2^2).$$

8. Find the volume generated by the revolution of the cycloid,

$$x = a(\psi - \sin\psi), \qquad y = a(1 - \cos\psi),$$

about its base. $5\pi^2a^3$.

9. The area included between the cycloid and tangents at the cusp and at the vertex revolves about the latter; find the volume generated.

π^2a^3.

10. Find the volume generated by the revolution of the part of the curve

$$y = \varepsilon^x,$$

which is on the left of the origin, about the axis of x.

$\frac{\pi}{2}$.

11. The axes of two equal right circular cylinders, whose common radius is a, intersect at the angle α; find the volume common to the cylinders.

The section parallel to the axes is a rhombus. $\dfrac{16a^3}{3\sin\alpha}$.

12. Find the volume generated by the revolution of one branch of the sinusoid,

$$y = b\sin\frac{x}{a},$$

about the axis of x. $\dfrac{\pi^2 b}{2a}$.

13. Find the volume enclosed by the surface generated by the revolution of an arc of a parabola about a chord, whose length is $2c$, perpendicular to the axis, and at a distance b from the vertex.

$\dfrac{16\pi b^2 c}{15}$.

14. Find the volume generated by the revolution of the tractrix, whose differential equation is

$$\frac{dy}{dx} = \pm\frac{y}{\sqrt{(a^2 - y^2)}},$$

about the axis of x.

Express $\pi y^2\,dx$ in terms of y. $\dfrac{2\pi a^3}{3}$.

15. Find the volume cut from a right circular cylinder whose radius is a, by a plane passing through the centre of the base, and making the angle α with the plane of the base.

$\dfrac{2a^3\tan\alpha}{3}$.

16. Find the volume generated by the curve

$$xy^2 = 4a^2(2a - x)$$

revolving about its asymptote. $4\pi^2 a^3$.

17. Express the volume of a frustum of a cone in terms of its height h, and the radii a_1 and a_2 of its bases.

$$\frac{\pi h}{3}(a_1^2 + a_1 a_2 + a_2^2).$$

18. Find the volume generated by the revolution of the cardioid,

$$r = a(1 - \cos\theta),$$

about the initial line.

Express y and dx in terms of θ.

$$\frac{8\pi a^3}{3}.$$

19. Find the volume of a barrel whose height is $2h$, and diameter $2b$, the longitudinal section through the centre being a segment of an ellipse whose foci are in the ends of the barrel.

$$2\pi b^2 h \frac{2h^2 + 3b^2}{3(b^2 + h^2)}.$$

20. Find the volume generated by the superior and by the inferior branch of the conchoid each revolving about the directrix; the equation, when the axis of y is the directrix, being

$$x^2 y^2 = (a + x)^2 (b^2 - x^2).$$

$$\pi^2 ab^2 \pm \frac{4\pi b^3}{3}.$$

21. On two opposite lateral faces of a rectangular parallelopiped whose base is ab, oblique lines are drawn, cutting off the distances c_1, c_2, c_3, c_4 on the lateral edges. A straight line intersecting each of these lines moves across the parallelopiped, remaining always parallel to the other lateral faces: find the volume cut off.

$$\frac{ab(c_1 + c_2 + c_3 + c_4)}{4}.$$

22. Find the volume enclosed by the surface generated by an arc of a circle whose radius is a, about a chord whose length is $2c$.

$$\frac{2\pi c(3a^2 - c^2)}{3} - 2\pi a^2 \sqrt{(a^2 - b^2)} \sin^{-1}\frac{c}{a}.$$

23. The area included between a quadrant of the ellipse

$$x = a\cos\phi, \qquad y = b\sin\phi,$$

and the tangents at its extremities revolves about the tangent at the extremity of the minor axis; find the volume generated.

$$\frac{\pi ab^2(10 - 3\pi)}{6}.$$

24. An ellipse revolves about the tangent at the extremity of its major axis; express the entire volume in the form of an integral, whose limits are o and 2π, and find its value. $2\pi^2a^2b$.

25. Show that the volume between the surface,

$$z^n = a^2x^2 + b^2y^2,$$

and any plane parallel to the plane of xy is equal to the circumscribing cylinder divided by $n + 1$.

26. A straight line of fixed length $2c$ moves with its extremities in two fixed perpendicular straight lines not in the same plane, and at a distance $2b$. Prove that every point in the moving line describes an ellipse in a plane parallel to both the fixed lines, and find the volume enclosed by the generated surface.

$$\frac{4\pi(c^2 - b^2)b}{3}.$$

27. Find the volume enclosed by the surface whose equation is

$$\frac{x^2}{a^2} + \frac{y^2}{b^2} + \frac{z^4}{c^4} = 1. \qquad \frac{8\pi abc}{5}.$$

28. A moving straight line, which is always perpendicular to a fixed straight line through which it passes, passes also through the circumference of a circle whose radius is a, in a plane parallel to the fixed straight line and at a distance b from it; find the volume enclosed by the surface generated and the circle.

$$\frac{\pi a^2b}{2}.$$

29. Find the volume enclosed by the surface

$$\frac{y^2}{b^2} + \frac{z^2}{c^2} = \frac{x}{a}$$

and the plane $x = a$. $\frac{\pi abc}{2}$.

30. Find the volume enclosed by the surface

$$x^{\frac{2}{3}} + y^{\frac{2}{3}} + z^{\frac{2}{3}} = a^{\frac{2}{3}}.$$

Find A_z as in Art. 107, *and then evaluate V by a similar method.*

$\frac{4\pi a^3}{35}$.

31. Find the volume between the coordinate planes and the surface

$$\left(\frac{x}{a}\right)^{\frac{1}{2}} + \left(\frac{y}{b}\right)^{\frac{1}{2}} + \left(\frac{z}{c}\right)^{\frac{1}{2}} = 1.$$ $\frac{abc}{90}$.

32. Find the volume cut from the paraboloid of revolution

$$y^2 + z^2 = 4ax$$

by the right circular cylinder

$$x^2 + y^2 = 2ax,$$

whose axis intersects the axis of the paraboloid perpendicularly at the focus, and whose surface passes through the vertex. $2\pi a^3 + \frac{16a^3}{3}$.

33. The paraboloid of revolution

$$x^2 + y^2 = cz$$

is pierced by the right circular cylinder

$$x^2 + y^2 = ax,$$

whose diameter is a, and whose surface contains the axis of the paraboloid; find the volume between the plane of xy and the surfaces of the paraboloid and of the cylinder. $\dfrac{3\pi a^4}{32c}$.

34. Find the volume cut from a sphere whose radius is a by a right circular cylinder whose radius is b, and whose axis passes through the centre of the sphere. $\dfrac{4\pi}{3}\left[a^3 - (a^2 - b^2)^{\frac{3}{2}}\right]$.

35. Find the volume cut from a sphere whose radius is a by the cylinder whose base is the curve

$$r = a \cos 3\theta. \qquad \frac{2a^3\pi}{3} - \frac{8a^3}{9}.$$

36. Find the volume cut from a sphere whose radius is a by the cylinder whose base is the curve

$$r^2 = a^2 \cos^2 \theta + b^2 \sin^2 \theta,$$

supposing $b < a$. $\dfrac{4\pi a^3}{3} - \dfrac{16}{9}(a^2 - b^2)^{\frac{3}{2}}$.

37. A right cone, the radius of whose base is a and whose altitude is b, is pierced by a cylinder whose base is a circle having for diameter a radius of the base of the cone; find the volume common to the cone and the cylinder. $\dfrac{ba^2}{36}(9\pi - 16)$.

38. The axis of a right cone whose semi-vertical angle is α coincides with a diameter of the sphere whose radius is a, the vertex being on the surface of the sphere; find the volume of the portion of the sphere which is outside of the cone. $\dfrac{4\pi a^3 \cos^4 \alpha}{3}$.

39. Find the volume produced by the revolution of the lemniscata

$$r^2 = a^2 \cos 2\theta,$$

about a perpendicular to the initial line. $\dfrac{\pi^2 a^3 \sqrt{2}}{8}$

40. Find the volumes generated by the revolution of the large loop and by one of the small loops of the curve

$$r = a \cos\theta \cos 2\theta$$

about a perpendicular to the initial line.

$$\frac{\pi^2 a^3}{16} + \frac{\pi a^3}{5}, \text{ and } \frac{\pi^2 a^3}{32} - \frac{\pi a^3}{10}.$$

41. From the element

$$r\,dr\,d\theta\,dz$$

derive the formulas for determining the volume of a solid of revolution whose axis is the axis of z.

$$V = 2\pi \iint r\,dr\,dz,$$

$$V = \pi \int (r_2^2 - r_1^2)\,dz, \qquad \text{and} \qquad V = 2\pi \int (z_2 - z_1) r\,dr.$$

Interpret the elements in these integrals.

42. Find the volume generated by the revolution of the curve

$$(x^2 + y^2)^2 = a^2x^2 + b^2y^2,$$

in which $a > b$, about the axis of y.

Transform to polar coordinates, and use the method of Art. 139.

$$\frac{\pi b(2b^2 + 3a^2)}{6} + \frac{\pi a^4}{2\sqrt{(a^2 - b^2)}} \cos^{-1}\frac{b}{a}.$$

43. Find the volume generated by the curve given in the preceding example, when revolving about the axis of x.

$$\frac{\pi a(2a^2 + 3b^2)}{6} + \frac{\pi b^4}{2\sqrt{(a^2 - b^2)}} \cdot \log\frac{a + \sqrt{(a^2 - b^2)}}{b}.$$

44. Find the volume common to the sphere whose radius is $\rho = a$, and to the solid formed by the revolution of the cardioid,

$$r = a(1 + \cos\theta),$$

about the initial line.

See Art. 141. $\dfrac{5\pi a^3}{6}$.

45. Find the whole volume enclosed by the surface

$$(x^2 + y^2 + z^2)^3 = a^3 xyz.$$

Transform to the coordinates ρ, ϕ, θ, *and show that the solid consists of four equal detached parts.* $\dfrac{a^3}{6}$.

X.

Rectification of Plane Curves.

142. A curve is said to be *rectified* when its length is determined, the unit of measure to which it is referred being a right line.

It is shown in Diff. Calc., Art. 314 [Abridged Ed., Art. 164], that, if s denotes the length of the arc of a curve given in rectangular coordinates, we shall have

$$ds = \sqrt{(dx^2 + dy^2)}.$$

If the abscissas of the extremities of the arc are known, s is found by substituting for dy in this expression its value in terms of x and dx, and integrating the result between the given values of x as limits. Thus, to express the arc measured from the vertex of the semi-cubical parabola

$$ay^2 = x^3$$

in terms of the abscissa of its other extremity, we derive, from the equation of the curve,

$$dy = \frac{3\sqrt{x}\,dx}{2\sqrt{a}},$$

whence $$ds = \frac{\sqrt{(9x+4a)}}{2\sqrt{a}}\,dx.$$

Integrating,

$$s = \frac{1}{2\sqrt{a}}\int_0 \sqrt{(9x + 4a)}\,dx$$

$$= \frac{1}{27\sqrt{a}}(9x + 4a)^{\frac{3}{2}} - \frac{8a}{27}.$$

143. When x and y are given in terms of a third variable, ds is generally expressed in terms of this variable. For example, from the equations of the four-cusped hypocycloid,

$$x = a\cos^3\psi, \qquad y = a\sin^3\psi, \quad . \quad . \quad . \quad (1)$$

we derive

$$dx = -\,3a\cos^2\psi\sin\psi\,d\psi, \quad \text{and} \quad dy = 3a\sin^2\psi\cos\psi\,d\psi;$$

whence $$ds = 3a\sin\psi\cos\psi\,d\psi. \quad . \quad . \quad . \quad . \quad . \quad . \quad (2)$$

The length of the arc between the point $(a, 0)$, corresponding to $\psi = 0$, and $(0, a)$ corresponding to $\psi = \frac{1}{2}\pi$, is therefore

$$\frac{3a}{2}\sin^2\psi\Big]_0^{\frac{\pi}{2}} = \frac{3a}{2}.$$

Change of the Sign of ds.

144. We have hitherto assumed ds to be positive, but it is to be remarked that an expression substituted for ds, as in the illustration given in the preceding article, may change sign. Thus, in equation (2), ds, which is so written as to be positive while ψ passes from 0 to $\frac{1}{2}\pi$, becomes negative while ψ passes from $\frac{1}{2}\pi$ to π. Thus the integral gives a negative result for the arc between the points $(0, a)$ and $(-a, 0)$, corresponding to $\frac{1}{2}\pi$ and π. This change of sign in ds indicates a *cusp* or *stationary point* of the curve; and the existence of such points must be considered before we can properly interpret the resulting values of s. For instance, if in this example we integrate between the limits 0 and $\frac{3\pi}{4}$, we get the result $s = \frac{3a}{4}$, which is *the algebraic sum*, but *the numerical difference* of the arcs between the points corresponding to the limits.

Polar Coordinates.

145. It is proved in Diff. Calc., Art. 317 [Abridged Ed., Art. 167], that when the curve is given in polar coordinates

$$ds = \sqrt{(dr^2 + r^2\,d\theta^2)}.$$

This is usually expressed in terms of θ. For example, the equation of the cardioid is

$$r = a(1 - \cos\theta) = 2a \sin^2 \tfrac{1}{2}\theta;$$

whence

$$dr = 2a \sin \tfrac{1}{2}\theta \cos \tfrac{1}{2}\theta\, d\theta,$$

and by substitution

$$ds = 2a \sin \tfrac{1}{2}\theta\, d\theta.$$

The limits for the whole perimeter of the curve are 0 and 2π, and ds remains positive for the whole interval. Therefore

$$s = 2a\int_0^{2\pi} \sin\frac{\theta}{2}\,d\theta = -4a\cos\frac{\theta}{2}\Big]_0^{2\pi} = 8a.$$

Rectification of Curves of Double Curvature.

146. Let σ denote the length of the arc of *a curve of double curvature*; that is, one which does not lie in a plane, and suppose the curve to be referred to rectangular coordinates x, y and z. If at any point of the curve the differentials of the coordinates be drawn in the directions of their respective axes, a rectangular parallelopiped will be formed, whose sides are dx, dy and dz, and whose diagonal is $d\sigma$. Hence

$$d\sigma = \sqrt{(dx^2 + dy^2 + dz^2)}.$$

The curve is determined by means of two equations connecting x, y and z, one of which usually expresses the value of y in terms of x, and the other that of z in terms of x. We can then express $d\sigma$ in terms of x and dx.

If the given equations contain all the variables, equations of the required form may be obtained by elimination.

147. An equation containing the two variables x and y only is evidently the equation of *the projection upon the plane of xy* of a curve traced upon the surface determined by the other equation. Let s denote the length of this projection; then, since $ds^2 = dx^2 + dy^2$,

$$d\sigma = \sqrt{(ds^2 + dz^2)},$$

in which ds may, if convenient, be expressed in polar coordinates; thus,

$$d\sigma = \sqrt{(dr^2 + r^2\,d\theta^2 + dz^2)}.$$

148. As an illustration, let us use this formula to determine the length of the loxodromic curve from the equation of the sphere,

$$x^2 + y^2 + z^2 = a^2, \quad . \quad . \quad . \quad . \quad . \quad . \quad . \quad . \quad (1)$$

upon which it is traced, and its projection upon the plane of the equator, of which the equation is

$$2a = \sqrt{(x^2 + y^2)}\left(\epsilon^{n \tan^{-1}\frac{y}{x}} + \epsilon^{-n \tan^{-1}\frac{y}{x}}\right),$$

or in polar coordinates

$$2a = r\,(\epsilon^{n\theta} + \epsilon^{-n\theta}). \quad . \quad . \quad . \quad . \quad . \quad . \quad . \quad (2)$$

Equation (1) is equivalent to

$$r^2 + z^2 = a^2;$$

and, denoting the latitude of the projected point by ϕ, this gives

$$z = a \sin\phi, \qquad\qquad r = a\cos\phi. \quad . \quad . \quad . \quad (3)$$

In order to express $d\theta$ in terms of ϕ, we substitute the value of r in (2); whence

$$\epsilon^{n\theta} + \epsilon^{-n\theta} = 2\sec\phi, \quad . \quad . \quad . \quad . \quad . \quad . \quad . \quad (4)$$

and by differentiation

$$\epsilon^{n\theta} - \epsilon^{-n\theta} = \frac{2}{n}\sec\phi\tan\phi\frac{d\phi}{d\theta}. \quad . \quad . \quad . \quad . \quad . \quad (5)$$

Squaring and subtracting equation (5) from equation (4),

$$4 = \frac{4\sec^2\phi}{n^2}\left[n^2 - \tan^2\phi\frac{d\phi^2}{d\theta^2}\right],$$

which reduces to

$$d\theta^2 = \frac{\sec^2\phi\,d\phi^2}{n^2}. \quad . \quad . \quad . \quad . \quad . \quad . \quad . \quad . \quad (6)$$

From equations (3) and (6)

$$r^2\,d\theta^2 = \frac{a^2}{n^2}\,d\phi^2,$$

$$dr^2 = a^2 \sin^2\phi\, d\phi^2,$$
$$dz^2 = a^2 \cos^2\phi\, d\phi^2;$$

whence substituting in the value of $d\sigma$ (p. 171)

$$d\sigma = a\sqrt{\left(1 + \frac{1}{n^2}\right)}\, d\phi.$$

Integrating,

$$\sigma = a\frac{\sqrt{(n^2+1)}}{n}\int_\alpha^\beta d\phi = a\frac{\sqrt{(n^2+1)}}{n}(\beta - \alpha),$$

where α and β denote the latitudes of the extremities of the arc.

Examples X.

1. Find the length of an arc measured from the vertex of the catenary

$$y = \frac{c}{2}\left(\varepsilon^{\frac{x}{c}} + \varepsilon^{-\frac{x}{c}}\right),$$

and show that the area between the coordinate axes and any arc is proportional to the arc.

$$s = \frac{c}{2}\left(\varepsilon^{\frac{x}{c}} - \varepsilon^{-\frac{x}{c}}\right).$$
$$A = cs.$$

2. Find the length of an arc measured from the vertex of the parabola

$$y^2 = 4ax.$$

$$\sqrt{(ax + x^2)} + a \log \frac{\sqrt{x} + \sqrt{(x+a)}}{\sqrt{a}}.$$

3. Find the length of the curve

$$y = \frac{\varepsilon^x + 1}{\varepsilon^x - 1},$$

between the points whose abscissas are a and b.

$$\log \frac{\varepsilon^{2b} - 1}{\varepsilon^{2a} - 1} + a - b.$$

4. Find the length, measured from the origin, of the curve

$$y = a \log \frac{a^2 - x^2}{a^2}.$$

$$a \log \frac{a + x}{a - x} - x.$$

5. Given the differential equation of the tractrix,

$$\frac{dy}{dx} = - \frac{y}{\sqrt{(a^2 - y^2)}},$$

and, assuming (0, a) to be a point of the curve, find the value of s as measured from this point, and also the value of x in terms of y; that is, find the rectangular equation of the curve.

$$s = a \log \frac{y}{a}.$$

$$x = a \log \frac{a + \sqrt{(a^2 - y^2)}}{y} - \sqrt{(a^2 - y^2)}.$$

6. Find the length of one branch of the cycloid

$$x = a(\psi - \sin \psi), \qquad y = a(1 - \cos \psi).$$

$8a$.

7. When the cycloid is referred to its vertex, the equations being

$$x = a(1 - \cos \psi), \qquad y = a(\psi + \sin \psi),$$

prove that $$s = \sqrt{(8ax)}.$$

8. Find the length from the point $(a, 0)$ of the curve

$$x = 2a \cos \psi - a \cos 2\psi,$$

$$y = 2a \sin \psi - a \sin 2\psi.$$

$4a (\psi - \sin \psi)$.

9. Show that the curve,

$$x = 3a \cos \psi - 2a \cos^3 \psi, \qquad y = 2a \sin^3 \psi,$$

has cusps at the points given by $\psi = 0$ and $\psi = \pi$; and find the whole length of the curve. $12a$.

10. Find the length of a quadrant of the curve

$$\left(\frac{x}{a}\right)^{\frac{2}{3}} + \left(\frac{y}{b}\right)^{\frac{2}{3}} = 1.$$

See Fig. 6, *Art.* 107. $\dfrac{a^2 + ab + b^2}{a + b}$.

11. Show that the curve

$$x = 2a \cos^2 \theta (3 - 2 \cos^2 \theta), \qquad y = 4a \sin \theta \cos^3 \theta$$

has three cusps, and that the length of each branch is $\dfrac{8a}{3}$.

12. Find the length of the arc between the points at which the curve

$$x = a \cos^2 \theta \cos 2\theta, \qquad y = a \sin^2 \theta \sin 2\theta$$

cuts the axes. $\dfrac{2 - \sqrt{2}}{3} a$.

13. Show that the curve

$$x = a \cos \psi (1 + \sin^2 \psi),$$
$$y = a \sin \psi \cos^2 \psi$$

is symmetrical to the axes, and find the length of the arcs between the cusps.

$$a\left(\sqrt{2} - \sin^{-1}\frac{1}{\sqrt{3}}\right);$$

$$a\left(\sqrt{2} + \cos^{-1}\frac{1}{\sqrt{3}}\right).$$

14. Find the length of one branch of the epicycloid

$$x = (a + b) \cos \psi - b \cos \frac{a + b}{b} \psi,$$

$$y = (a + b) \sin \psi - b \sin \frac{a + b}{b} \psi.$$

$$\frac{8b(a + b)}{a}.$$

15. Show that the curve

$$x = 9a \sin \psi - 4a \sin^3 \psi,$$
$$y = -3a \cos \psi + 4a \cos^3 \psi$$

is symmetrical to the axes, and has double points and cusps: find the lengths of the arcs, (α) between the double points, (β) between a double point and a cusp, and (γ) the arc connecting two cusps, and not passing through the double points.

(α), $a(\pi + 3\sqrt{3})$;

(β), $\frac{\pi a}{2}$;

(γ), $a(3\sqrt{3} - \pi)$.

16. Find the whole length of the curve

$$x = 3a \sin \psi - a \sin^3 \psi,$$
$$y = a \cos^3 \psi.$$

$3\pi a$.

17. Find the length, measured from the pole, of any arc of the equiangular spiral

$$r = a\varepsilon^{n\theta},$$

in which $n = \cot\alpha$. $r \sec\alpha$.

18. Prove by integration that the arc subtending the angle θ at the circumference in a circle whose radius is a, is $2a\theta$.

19. Find the length, measured from the origin, of the curve defined by the equations

$$y = \frac{x^2}{2a}, \qquad z = \frac{x^3}{6a^2}.$$

$$x + \frac{x^3}{6a^2}.$$

20. Find the length, measured from the origin, of the intersection of the surfaces

$$y = 4n \sin x, \qquad z = 2n^2 (2x + \sin 2x).$$

$$(4n^2 + 1)x + 2n^2 \sin 2x.$$

21. Find the length, measured from the origin, of the intersection of the cylindrical surfaces

$$(y - x)^2 = 4ax, \qquad 9a(z - x)^2 = 4x^3.$$

$$\frac{2x^{\frac{3}{2}}}{3\sqrt{a}} + 2\sqrt{(ax)} + x.$$

22. If upon the hyperbolic cylinder

$$\frac{y^2}{c^2} - \frac{z^2}{b^2} = 1,$$

a curve whose projection upon the plane of xy is the catenary

$$y = \frac{c}{2}\left(\varepsilon^{\frac{x}{c}} + \varepsilon^{-\frac{x}{c}}\right)$$

be traced, prove that any arc of the curve bears to the corresponding arc of its projection the constant ratio $\sqrt{(b^2 + c^2)} : c$.

XI.

Surfaces of Solids of Revolution.

149. The surface of a solid of revolution may be generated by the circumference of the circular section made by a plane perpendicular to the axis of revolution. Thus in Fig. 17, the surface produced by the revolution of the curve AB about the axis of x is regarded as generated by the circumference PQ. The radius of this circumference is y, and its *plane* has a motion whose differential is dx, but every point in the circumference itself has a motion whose differential is ds, s denoting an arc of the curve AB.

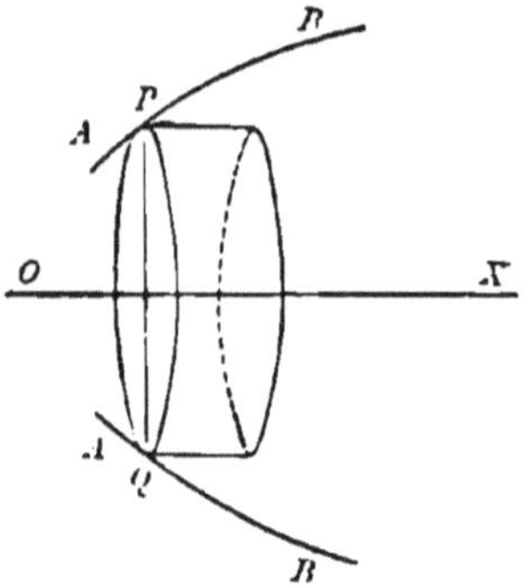

FIG. 17.

Hence, denoting the required surface by S, we have

$$dS = 2\pi y\, ds = 2\pi y \sqrt{(dx^2 + dy^2)}.$$

The value of dS must of course be expressed in terms of a single variable before integration.

150. For example, let us determine the area of the zone of spherical surface included between any two parallel planes. The radius of the sphere being a, the equation of the revolving curve is

$$x^2 + y^2 = a^2;$$

whence

$$y = \sqrt{(a^2 - x^2)},$$

$$dy = -\frac{x\, dx}{\sqrt{(a^2 - x^2)}},$$

$$ds = \frac{a\, dx}{\sqrt{(a^2 - x^2)}},$$

and

$$dS = 2\pi a\, dx;$$

therefore

$$S = 2\pi a \int dx = 2\pi a\,(x_2 - x_1).$$

Since $x_2 - x_1$ is the distance between the parallel planes, the area of a zone is the product of its altitude by $2\pi a$, the circumference of a great circle, and the area of the whole surface of the sphere is $4\pi a^2$.

151. When the curve is given in polar coordinates, it is convenient to transform the expression for S to polar coordinates. Thus, if the curve revolves about the initial line,

$$S = 2\pi \int y\,ds = 2\pi \int r \sin\theta \sqrt{(dr^2 + r^2\,d\theta^2)}.$$

For example, if the curve is the cardioid

$$r = 2a \sin^2 \frac{1}{2}\theta,$$

we find, as in Art. 145,

$$ds = 2a \sin \frac{1}{2}\theta\,d\theta.$$

Hence

$$S = 16\pi a^2 \int_0^\pi \sin^4 \frac{1}{2}\theta \cos \frac{1}{2}\theta\,d\theta$$

$$= \frac{32\pi a^2}{5} \sin^5 \frac{1}{2}\theta \Big]_0^\pi = \frac{32\pi a^2}{5}.$$

Areas of Surfaces in General.

152. Let a surface be referred to rectangular coordinates x, y and z; the projection of a given portion of the surface upon the plane of xy is a plane area determined by a given relation between x and y. We may take as the elements of the surface the portions which are projected upon the corresponding elements of area in the plane of xy. If at a point within the element of surface, which is projected upon a given element $\Delta x \Delta y$, a tangent plane be passed, and if γ denote the inclination of this plane to the plane of xy, the area of the corresponding element *in the tangent plane* is

$$\sec \gamma \; \Delta x \Delta y.$$

The surface is evidently the limit of the sum of the elements in the tangent planes when Δx and Δy are indefinitely diminished. Now $\sec \gamma$ is a function of the coordinates of the point of contact of the tangent plane; and since these coordinates are values of x and y which lie respectively between x and $x + \Delta x$ and between y and $y + \Delta y$, the theorem proved in Art. 99 shows that this limit is

$$S = \iint \sec \gamma \, dx \, dy.$$

153. The value of $\sec \gamma$ may be derived by the following method. Through the point P of the surface let planes be passed parallel to the coordinate planes, and let PD, and PE, Fig. 17, be the intersections of the tangent plane with the planes parallel to the planes of xz and yz. Then PD and PE are tangents at P to the sections of the surface made by these planes. The equations of these sections are found by regarding y and x in turn as constants in the equation of the surface; therefore denoting the inclinations of these tangent lines to the plane of xy by ϕ and ψ, we have

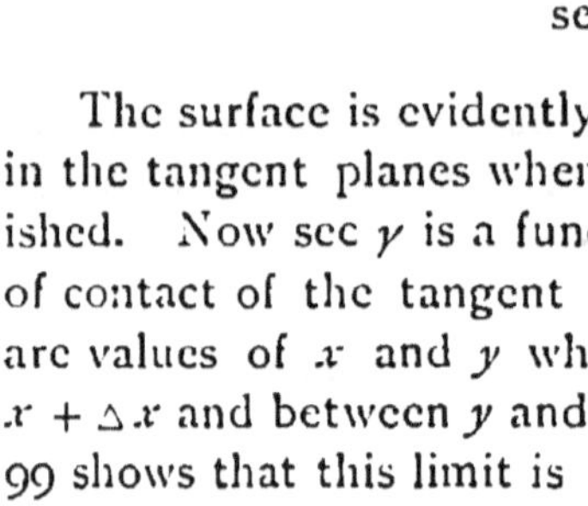

FIG. 18.

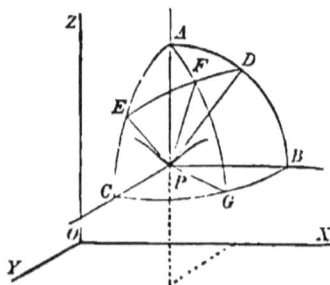

$$\tan \phi = \frac{dz}{dx}, \qquad \text{and} \qquad \tan \psi = \frac{dz}{dy},$$

in which $\frac{dz}{dx}$ and $\frac{dz}{dy}$ are partial derivatives derived from the equation of the surface.

If the planes be intersected by a spherical surface whose centre is P, ADE is a spherical triangle right angled at A, whose sides are the complements of ϕ and ψ. Moreover, if a plane perpendicular to the tangent plane PED be passed through AP, the angle FPG will be γ, and the perpendicular from the right angle to the base of the triangle the complement of γ.

Denoting the angle EAF by θ, the formulas for solving spherical right triangles give

$$\cos\theta = \frac{\tan\psi}{\tan\gamma}, \qquad \text{and} \qquad \sin\theta = \frac{\tan\phi}{\tan\gamma}.$$

Squaring and adding,

$$1 = \frac{\tan^2\psi + \tan^2\phi}{\tan^2\gamma},$$

$$\text{or } \tan^2\gamma = \tan^2\psi + \tan^2\phi;$$

whence $$\sec^2\gamma = 1 + \left(\frac{dz}{dx}\right)^2 + \left(\frac{dz}{dy}\right)^2.$$

Substituting in the formula derived in Art. (152), we have

$$S = \iint \sqrt{\left[1 + \left(\frac{dz}{dx}\right)^2 + \left(\frac{dz}{dy}\right)^2\right]}\, dx\, dy.$$

154. It is sometimes more convenient to employ the polar element of the projected area. Thus the formula becomes

$$S = \iint \sec \gamma \, r \, dr \, d\theta,$$

where sec γ has the same meaning as before.

For example, let it be required to find the area of the surface of a hemisphere intercepted by a right cylinder having a radius of the hemisphere for one of its diameters. From the equation of the sphere,

$$x^2 + y^2 + z^2 = a^2, \quad \dots \dots \dots \quad (1)$$

we derive

$$\frac{dz}{dx} = -\frac{x}{z}, \qquad \frac{dz}{dy} = -\frac{y}{z};$$

whence

$$\sec \gamma = \sqrt{\left[1 + \sqrt{\left(\frac{dz}{dx}\right)^2 + \left(\frac{dz}{dy}\right)^2}\right]} = \frac{a}{z};$$

therefore

$$S = a \iint \frac{r \, dr \, d\theta}{z},$$

the integration extending over the area of the circle

$$r = a \cos \theta. \quad \dots \dots \dots \quad (2)$$

Since equation (1) is equivalent to

$$z^2 + r^2 = a^2,$$

$$S = a \iint_{r_1}^{r_2} \frac{r \, dr}{\sqrt{(a^2 - r^2)}} \, d\theta = a \int [\sqrt{(a^2 - r_1^2)} - \sqrt{(a^2 - r_2^2)}] \, d\theta.$$

From (2) the limits for r are $r_1 = 0$, and $r_2 = a\cos\theta$, hence

$$S = a^2\int(1 - \sin\theta)\,d\theta,$$

in which $a\sin\theta$ is put for the *positive* quantity $\sqrt{(a^2 - r_2^2)}$. The limits for θ are $-\frac{1}{2}\pi$ and $\frac{1}{2}\pi$, but since $\sin\theta$ is in this case to be regarded as invariable in sign, we must write

$$S = 2a^2\int_0^{\frac{\pi}{2}}(1 - \sin\theta)\,d\theta = \pi a^2 - 2a^2.$$

If another cylinder be constructed, having the opposite radius of the hemisphere for diameter, the surface removed is $2\pi a^2 - 4a^2$, and the surface which remains is $4a^2$, a quantity commensurable with the square of the radius. This problem was proposed in 1692, in the form of an enigma, by Viviani, a Florentine mathematician.

Examples XI.

1. Find the surface of the paraboloid whose altitude is a, and the radius of whose base is b.

$$\frac{\pi b}{6a^2}\left[(4a^2 + b^2)^{\frac{3}{2}} - b^3\right]$$

2. Prove that the surface generated by the arc of the catenary given in Ex. X., 1, revolving about the axis of x, is equal to

$$\pi(cx + sy).$$

3. Find the whole surface of the oblate spheroid produced by the

revolution of an ellipse about its minor axis, a denoting the major, b the minor semi-axis, and e the excentricity, $\frac{\sqrt{(a^2 - b^2)}}{a}$.

$$2\pi a^2 + \pi \frac{b^2}{e} \log \frac{1 + e}{1 - e}.$$

4. Find the whole surface of the prolate spheroid produced by the revolution of the ellipse about its major axis, using the same notation as in Ex. 3.

$$2\pi b^2 + 2\pi ab \frac{\sin^{-1} e}{e}.$$

5. Find the surface generated by the cycloid

$$x = a(\psi - \sin\psi), \qquad y = a(1 - \cos\psi)$$

revolving about its base. $\frac{64}{3}\pi a^2$.

6. Find the surface generated when the cycloid revolves about the tangent at its vertex.

$$\frac{32}{3}\pi a^2.$$

7. Find the surface generated when the cycloid revolves about its axis.

$$8\pi a^2\left(\pi - \frac{4}{3}\right).$$

8. Find the surface generated by the revolution of one branch of the tractrix (see Ex. X., 5) about its asymptote.

$$2\pi a^2.$$

9. Find the surface generated by the revolution about the axis of x of the portion of the curve

$$y = \varepsilon^x,$$

which is on the left of the axis of y.

$$\pi[\sqrt{2} + \log(1 + \sqrt{2})].$$

10. Find the surface generated by the revolution about the axis of x of the arc between the points for which $x = a$ and $x = b$ in the hyperbola

$$xy = k^2.$$

$$\pi k^2\left[\log\frac{b^2 + \sqrt{(k^4 + b^4)}}{a^2 + \sqrt{(k^4 + a^4)}} + \frac{\sqrt{(k^4 + a^4)}}{a^2} - \frac{\sqrt{(k^4 + b^4)}}{b^2}\right].$$

11. Show that the surface of a cylinder whose generating lines are parallel to the axis of z is represented by the integral

$$S = \int z\, ds,$$

where s denotes the arc of the base in the plane of xy. Hence, deduce the surface cut from a right circular cylinder whose radius is a, by a plane passing through the centre and making the angle α with the plane of the base. $2a^2 \tan\alpha$.

12. Find the surface of that portion of the cylinder in the problem solved in Art. 154, which is within the hemisphere. $2a^2$.

13. Find the surface of a circular spindle, a being the radius and $2c$ the chord.

$$4\pi a\left[c - \sqrt{(a^2 - c^2)}\sin^{-1}\frac{c}{a}\right].$$

XII.

The Area generated by a Straight Line moving in any Manner in a Plane.

155. If a straight line of indefinite length moves in any manner whatever in a plane, there is at each instant a point of the line about which it may be regarded as rotating. This point we shall call *the centre of rotation* for the instant. The rate of motion of every point of the line in a direction perpendicular to the line itself is at the instant the same as it would be if the line were rotating at the same angular rate about this point as a fixed centre.* Hence it follows that the area generated by a definite portion of the line has at the instant the same rate as if the line were rotating about a fixed instead of a variable centre.

156. Suppose at first that the centre of rotation is on the generating line produced, ρ_1 and ρ_2 denoting the distances from the centre of the extremities of the generating line, and let ϕ denote its inclination to a fixed line. By substitution in the general formula derived in Art. 110, we have

$$dA = \frac{1}{2}(\rho_2^2 - \rho_1^2)\, d\phi.$$

---

* Compare Diff. Calc., Art. 332 [Abridged Ed., Art. 176], where the moving line is the normal to a given curve, and the centre of rotation is the centre of curvature of the given curve. If the line is moving without change of direction, the centre is of course at an infinite distance.

When the line is regarded as forming a part of a rigidly connected system in motion, its centre of rotation is the foot of a perpendicular dropped upon it from the *instantaneous centre* of the motion of the system. Thus, if the tangent and normal in the illustration cited are rigidly connected, the centre of curvature, C, is the *instantaneous centre* of the motion of the system, and the point of contact, P, is the centre of rotation for the tangent line.

157. The area between a curve and its evolute may be generated by the radius of curvature ρ, whose inclination to the axis of x is $\phi + \frac{1}{2}\pi$, in which ϕ denotes the inclination of the tangent line. Since the centre of rotation is one extremity of the generating line ρ, the differential of this area is found by substituting in the general expression $\rho_1 = 0$ and $\rho_2 = 0$. Hence when ρ is expressed in terms of ϕ,

$$A = \frac{1}{2}\int \rho^2\, d\phi$$

expresses the area between an arc of a given curve, its evolute, and the radii of curvature of its extremities, the limits being the values of ϕ at the ends of the given arc.

158. For example, in the case of the cardioid

$$r = a(1 - \cos\theta),$$

it is readily shown, from the results obtained in Art. 145, that the angle between the tangent and the radius vector is $\frac{1}{2}\theta$; and therefore $\phi = \frac{3}{2}\theta$, and

$$\rho = \frac{ds}{d\phi} = \frac{4a}{3}\sin\frac{\phi}{3}.$$

To obtain the whole area between the curve and its evolute, the limits for θ are 0 and 2π; hence the limits for ϕ are 0 and 3π. Therefore

$$A = \frac{1}{2}\int_0^{3\pi} \rho^2\, d\phi = \frac{8a^2}{9}\int_0^{3\pi} \sin^2\frac{\phi}{3}\, d\phi = \frac{4\pi a^2}{3}.$$

159. As another application of the general formula of Art. 156, let one end of a line of fixed length a be moved along a given line in a horizontal plane, while a weight attached to the other extremity is drawn over the plane by the line, and is therefore always moving in the direction of the line itself. The line of fixed length in this case turns about the weight as a moving centre of rotation. Hence the area generated while the line turns through a given angle is the same as that of the corresponding sector of a circle whose radius is a.

The curve described by the weight is called *a tractrix*, and the line along which the other extremity is moved is *the directrix*. When the axis of x is the directrix, and the weight starts from the point $(0, a)$, the common tractrix is described; hence the area between this curve and the axis is $\frac{1}{4}\pi a^2$.

160. Again, in the generation of the cycloid, Diff. Calc., Art. 288 [Abridged Ed., Art. 156], the variable chord RP may be regarded as generating the area. The point R has a motion in the direction of the tangent RX; the point P partakes of this motion, which is the motion of the centre C, and also has an equal motion, due to the rotation of the circle in the direction of the tangent to the circle at P. Since the tangents at P and R are equally inclined to PR, the motion of P in a direction perpendicular to PR is double the component, in this direction, of the motion of R. Therefore the centre of rotation of PR is beyond R at a distance from it equal to PR. Hence, denoting PRO by ϕ,

$$\rho_1 = PR = 2a \sin \phi, \qquad \rho_2 = 2PR = 4a \sin \phi.$$

Substituting in the formula of Art. 156, we have for the area of the cycloid, since PRO varies from 0 to π,

$$A = 6a^2 \int_0^\pi \sin^2 \phi \, d\phi = 3\pi a^2.$$

Sign of the Generated Area.

161. Let AB be the generating line, and C the centre of rotation. The expression,

$$dA = \tfrac{1}{2}(\rho_2^2 - \rho_1^2)\, d\phi, \quad . \quad . \quad . \quad . \quad . \quad . \quad (1)$$

for the differential of the area, was obtained upon the supposition that A and B were on the same side of C. Then supposing $\rho_2 > \rho_1$, and that the line rotates in the positive direction, as in figure 19, *the differential of the area is positive;* and we notice that every point in the area generated is swept over by the line AB, *the left hand side as we face in the direction AB preceding.*

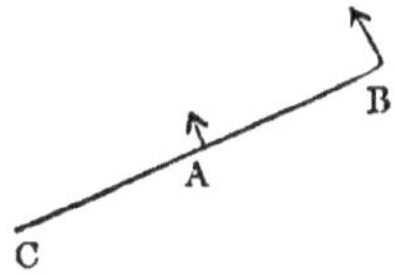

FIG. 19.

162. We shall now show that in every case, the formula requires that an area swept over with the left side preceding, shall be considered as positively generated, and one swept over in the opposite direction as negatively generated.

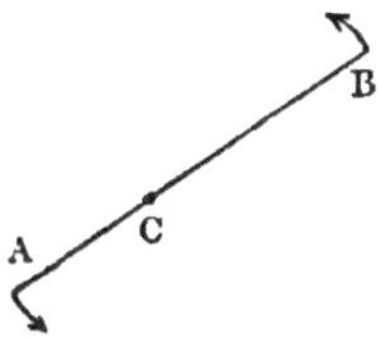

FIG. 20.

In the first place, if C is between A and B so that ρ_1 is negative, as in figure 20, ρ_1^2 is still positive, and formula (1) still gives the difference between the areas generated by AB and AC. Hence the latter area, which is now generated by a part of the line AB, must be regarded as generated negatively, but the *right hand side* as we face in the direction AB of this portion of the line is now preceding, which agrees with the rule given in Art. 161.

Again, if C is beyond B, the formula gives the difference of the generated areas; but since ρ_1^2 is numerically greater than ρ_2^2, in this case, dA is negative, and the area generated by AB is the difference of the areas, and is negative by the rule.

Finally, if the direction of rotation be reversed, $d\phi$ and therefore dA change sign, but the opposite side of each portion of the line becomes in this case the preceding side.

163. We may now put the expression for the area in another form. For

$$dA = \frac{1}{2}(\rho_2^2 - \rho_1^2)\,d\phi = (\rho_2 - \rho_1)\,\frac{\rho_2 + \rho_1}{2}\,d\phi;$$

whatever be the signs of ρ_2 and ρ_1, the first factor is the length of AB, which we shall denote by l, and the second factor is the distance of the middle point of AB from the centre of rotation, which we shall denote by ρ_m. Hence, putting

$$\rho_2 - \rho_1 = l, \qquad \text{and} \qquad \frac{\rho_2 + \rho_1}{2} = \rho_m,$$

we have

$$A = \int l\rho_m\,d\phi. \quad . \quad . \quad . \quad . \quad . \quad . \quad . \quad . \quad . \quad . \quad (2)$$

Since $\rho_m\,d\phi$ is the differential of the motion of the middle point in a direction perpendicular to AB, this expression shows that the differential of the area is the product of this differential by the length of the generating line.

Areas generated by Lines whose Extremities describe Closed Circuits.

164. Let us now suppose the generating line AB to move from a given position, and to return to the same position, each of the extremities A and B describing a closed curve in the positive direction, as indicated by the arrows in figure 21. It is readily seen that every point which is in the area described by B, and not in that described by A, will be swept over at least once by the line AB, the left side preceding, and if passed over more than once, there will be

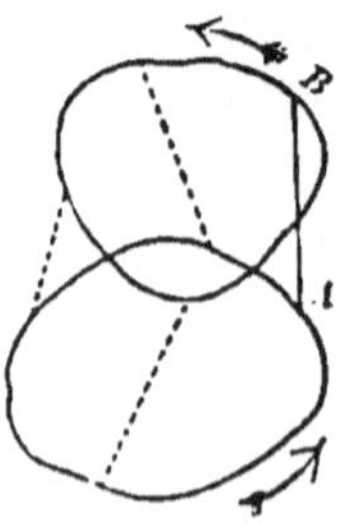

FIG. 21.

an excess of one passage, the left side preceding. Therefore the area within the curve described by B, and not within that described by A, will be generated positively. In like manner the area within the curve described by A, and not within that described by B, will be generated negatively. Furthermore, all points within both or neither of these curves are passed over, if at all, an equal number of times in each direction, so that the area common to the two curves and exterior to both disappears from the expression for the area generated by AB.

Hence it follows that, *regarding a closed area whose perimeter is described in the positive direction as positive, the area generated by a line returning to its original position is the difference of the areas described by its extremities.* This theorem is evidently true generally, if areas described in the opposite direction are regarded as negative.

Amsler's Planimeter.

165. The theorem established in the preceding article may be used to demonstrate the correctness of the method by which an area is measured by means of the *Polar Planimeter*, invented by Professor Amsler, of Schaffhausen.

This instrument consists of two bars, OA and AB, Fig. 22, jointed together at A. The rod OA turns on a fixed pivot at O, while a tracer at B is carried in the positive direction completely around the perimeter of the area to be measured. At some point C of the bar AB a small wheel is fixed, having its axis parallel to AB, and its circumference resting upon the paper. When B is moved, this wheel has a sliding and a rolling motion; the latter motion is recorded by an attachment by means of which the number of turns and parts of a turn of the wheel are registered.

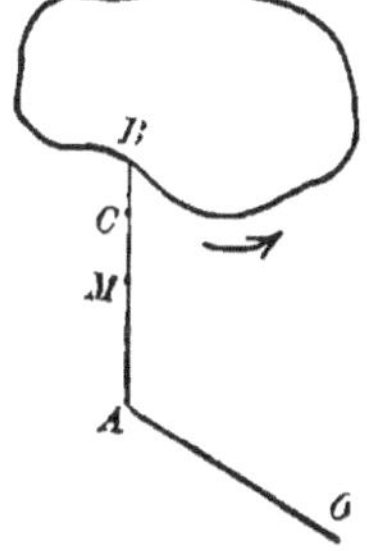

FIG. 22.

166. Let M be the middle point of AB, and let

$$OA = a, \qquad AB = b, \qquad MC = c.$$

Since b is constant, the area described by AB is by equation (2), Art. 163,

$$\text{Area } AB = b\int \rho_m \, d\phi. \quad . \quad . \quad . \quad . \quad . \quad . \quad . \quad (1)$$

Denoting the linear distance registered on the circumference of the wheel by s, ds is the differential of the motion of the point C, in a direction perpendicular to AB, and since the distance of this point from the centre of rotation is $\rho_m + c$,

$$ds = (\rho_m + c)\, d\phi:$$

substituting in (1) the value of $\rho_m \, d\phi$,

$$\text{Area } AB = b\int ds - bc\int d\phi. \quad . \quad . \quad . \quad . \quad . \quad (2)$$

167. Two cases arise in the use of the instrument. When, as represented in Fig. 22, O is outside the area to be measured, the point A describes no area, and by the theorem of Art. 164, equation (2) represents simply the area described by B. In this case ϕ returns to its original value, hence $\int d\phi$ vanishes, and denoting the area to be measured by A, equation (2) becomes

$$A = bs. \quad . \quad . \quad . \quad . \quad . \quad . \quad . \quad . \quad . \quad (3)$$

In the second case, when O is within the curve traced by B, the point A describes a circle whose area is πa^2, and the limiting values of ϕ differ by a complete revolution. Hence in this case equation (2) becomes

$$A - \pi a^2 = bs - 2\pi bc,$$

or

$$A = bs + \pi (a^2 - 2bc).* \quad . \quad . \quad . \quad . \quad . \quad (4)$$

In another form of the planimeter the point A moves in a straight line, and the same demonstration shows that the area is always equal to bs.

Examples XII.

1. The involute of a circle whose radius is a is drawn, and a tangent is drawn at the opposite end of the diameter which passes through the cusp; find the area between the tangent and the involute.

$$\frac{a^2\pi (3 + \pi^2)}{3}.$$

2. Two radii vectores of a closed oval are drawn from a fixed point within, one of which is parallel to the tangent at the extremity of the other; if the parallelogram be completed, the area of the locus of its vertex is double the area of the given oval.

3. Show that the area of the locus of the middle point of the chord joining the extremities of the radii vectores in Ex. 2, is one half the area of the given oval.

---

* The planimeter is usually so constructed that the positive direction of rotation is with the hands of a watch. The bar b is adjustable, but the distance AC is fixed so that c varies with b. Denoting AC by q, we have $c = q - \frac{1}{2}b$, and the constant to be added becomes $C = \pi (a^2 - 2bq + b^2)$ in which a and q are fixed and b adjustable. In some instruments q is negative.

It is to be noticed that in the second case s may be negative; the area is then the numerical difference between the constant and bs.

4. Prove that the difference of the perimeters of two parallel ovals, whose distance is b, is $2\pi b$, and that the difference of their areas is the product of b and the half sum of their perimeters.

5. A limaçon is formed by taking a fixed distance bc on the radius vector from a point on the circumference of a circle whose radius is a; show that the area generated by b when $b > 2a$ is the area of the limaçon diminished by twice the area of the circle, and thence determine the area of the limaçon.

$$\pi(2a^2 + b^2).$$

6. Verify equation (4), Art. 167, when the tracer describes the circle whose radius is $a + b$.

7. Verify the value of the constant in equation (4). Art. 167, by determining the circle which may be described by the tracer without motion of the wheel.

8. If, in the motion of a crank and connecting rod (the line of motion of the piston passing through the centre of the crank), Amsler's recording wheel be attached to the connecting rod at the piston end, determine s geometrically, and verify by means of the area described by the other end of the rod.

9. The length of the crank in Ex. 8 being a, and that of the connecting rod b, find the area of the locus of a point on the connecting rod at a distance c from the piston end.

$$\frac{\pi a^2 c}{b}.$$

10. If a line AB of fixed length move in a plane, returning to its original position *without making a complete revolution*, denoting the areas of the curves described by its extremities by (A) and (B), determine the area of the curve described by a point cutting AB in the ratio $m : n$.

$$\frac{n(A) + m(B)}{m + n}.$$

11. If the line in Ex. 10 return to its original position after *making a complete revolution*, prove *Holditch's Theorem*; namely, that the area of the curve described by a point at the distance c and c' from A and B is

$$\frac{c'(A) + c(B)}{c + c'} - \pi cc'.$$

12. Show by means of Ex. 11 that, if a chord of fixed length move around an oval, and a curve be described by a point at the distances c and c' from its ends, the area between the curves will be $\pi cc'$.

XIII.

Approximate Expressions for Areas and Volumes.

168. When the equation of a curve is unknown, the area between the curve, the axis of x, and two ordinates may be approximately expressed in terms of the base and a limited number of ordinates, which are supposed to have been measured.

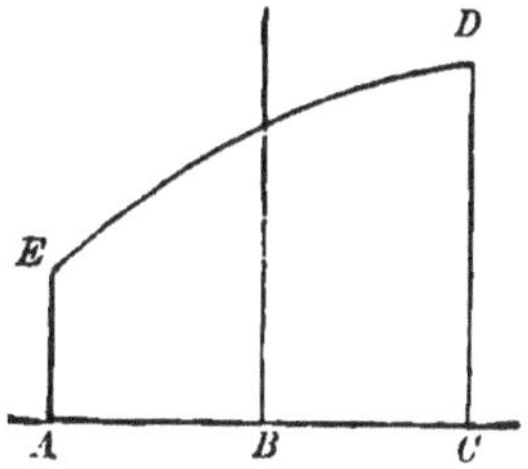

FIG. 23.

Let $ABCDE$ be the area to be determined; denote the length of the base by $2h$; and let the ordinates at the extremities and middle point of the base be measured and denoted by y_1, y_2, and y_3. Taking the base for the axis of x, and the middle point as origin, let it be assumed that the curve has an equation of the form

$$y = A + Bx + Cx^2 + Dx^3; \quad . \quad . \quad . \quad . \quad . \quad (1)$$

then the area required is

$$A = \int_{-h}^{h} y\,dx = Ax + \frac{Bx^2}{2} + \frac{Cx^3}{3} + \frac{Dx^4}{4}\Big]_{-h}^{h} = \frac{h}{3}(6A + 2Ch^2), \quad . \ (2)$$

in which which A and C are unknown.

In order to express the area in terms of the measured ordinates, we have from equation (1),

$$y_1 = A + Bh + Ch^2 + Dh^3,$$
$$y_2 = A,$$
$$y_3 = A - Bh + Ch^2 - Dh^3;$$

whence we derive

$$y_1 + y_3 = 2A + 2Ch^2,$$
$$y_1 + 4y_2 + y_3 = 6A + 2Ch^2;$$

and substituting in (2),

$$A = \frac{h}{3}(y_1 + 4y_2 + y_3).$$

It will be noticed that this formula gives a perfectly accurate result when the curve is really a parabolic curve of the third or a lower degree.

169. If the base be divided into three equal intervals, each denoted by h, and the ordinates at the extremities and at the points of division measured, we have, by assuming the same equation,

$$A = \int_{-\frac{3h}{2}}^{\frac{3h}{2}} y\,dx = \frac{3h}{4}(4A + 3Ch^2) \quad . \quad . \quad . \quad . \quad . \quad (1)$$

From the equation of the curve,

$$y_1 = A - \frac{3Bh}{2} + \frac{9Ch^2}{4} - \frac{27Dh^3}{8},$$

$$y_2 = A - \frac{Bh}{2} + \frac{Ch^2}{4} - \frac{Dh^3}{8},$$

$$y_3 = A + \frac{Bh}{2} + \frac{Ch^2}{4} + \frac{Dh^3}{8},$$

$$y_4 = A + \frac{3Bh}{2} + \frac{9Ch^2}{4} + \frac{27Dh^3}{8};$$

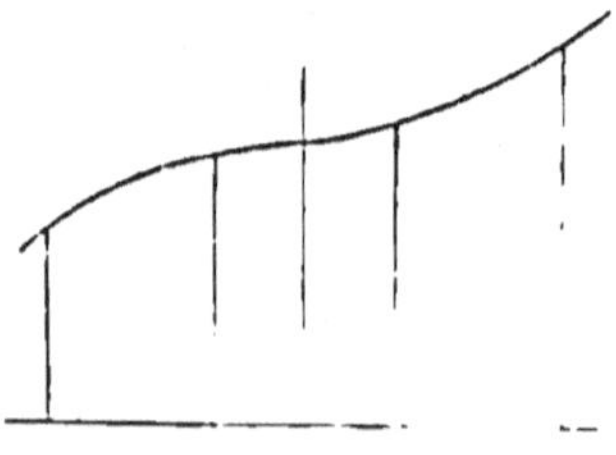

FIG. 24.

whence $$y'_1 + y'_4 = 2A + \frac{9Ch^2}{2},$$

$$y'_2 + y'_3 = 2A + \frac{Ch^2}{2}.$$

From these equations we obtain

$$A = \frac{-y'_1 + 9y'_2 + 9y'_3 - y'_4}{16},$$

and $$Ch^2 = \frac{y'_1 - y'_2 - y'_3 + y'_4}{4}.$$

Substituting in equation (1),

$$A = \frac{3h}{8}(y'_1 + 3y'_2 + 3y'_3 + y'_4).$$

Simpson's Rules.

170. The formulas derived in Articles 168 and 169, although they were first given by Cotes and Newton, are usually known as *Simpson's Rules*, the following extensions of the formulas having been published in 1743, in his *Mathematical Dissertations.*

If the whole base be divided into an even number n of parts, each equal to h, and the ordinates at the points of division be numbered in order from end to end, then by applying the first formula to the areas between the alternate ordinates, we have

$$A = \frac{h}{3}(y'_1 + 4y'_2 + 2y'_3 + 4y'_4 \cdots + 4y'_n + y'_{n+1}).$$

That is to say, the area is equal to the product of the sum of the extreme ordinates, four times the sum of the even-numbered ordinates, and twice the sum of the remaining odd-numbered ordinates, multiplied by one third of the common interval.

Again, if the base be divided into a number of parts divisible by three, we have, by applying the formula derived in Art. 169, to the areas between the ordinates $y_1 y_4$, $y_4 y_7$, and so on,

$$A = \frac{3h}{8}(y_1 + 3y_2 + 3y_3 + 2y_4 + 3y_5 \cdots + 3y_n + y_{n+1}).$$

Cotes' Method of Approximation.

171. The method employed in Articles 168 and 169 is known as *Cotes' Method.* It consists in assuming the given curve to be a parabolic curve of the highest order which can be made to pass through the extremities of a series of equidistant measured ordinates.

The equation of the parabolic curve of the nth order contains $n + 1$ unknown constants; hence, in order to eliminate these constants from the expression for an area defined by the curve, it is in general necessary to have $n + 1$ equations connecting them with the measured ordinates. Hence, if n denote the number of intervals between measured ordinates over which the curve extends, the curve will in general be of the nth degree.*

* If H denotes the whole base, the first factor is always equivalent to H divided by the sum of the coefficients of the ordinates; for if all the ordinates are made equal, the expression must reduce to Hy_1. Thus, each of the rules for an approximate area, including those derived by repeated applications, as in Art. 170, may be regarded as giving an expression for the *mean ordinate*. The coefficients of the ordinates, according to Cotes' method, for all values of n up to $n = 10$, may be found in Bertrand's *Calcul Intégral*, pages 333 and 334. For example (using detached coefficients for brevity), we have, when $n = 4$,

$$A = \frac{H}{90}[7, 32, 12, 32, 7];$$

and when $n = 6$,

$$A = \frac{H}{840}[41, 216, 27, 272, 27, 216, 41].$$

172. For example, let it be required to determine the area between the ordinates y_1 and y_2, in terms of the three equidistant ordinates y_1, y_2 and y_3, the common interval being h.

We must assume

$$y = A + Bx + Cx^2;$$

then taking the origin at the foot of y_1,

$$A = \int_0^h y\,dx = h\left[A + \frac{Bh}{2} + \frac{Ch^2}{3}\right],$$

from which A, B and C must be eliminated by means of the equations

$$y_1 = A,$$
$$y_2 = A + Bh + Ch^2,$$
$$y_3 = A + 2Bh + 4Ch^2.$$

Solving these equations, we obtain

$$A = y_1,$$
$$Bh = \frac{-3y_1 + 4y_2 - y_3}{2},$$
$$Ch^2 = \frac{y_1 - 2y_2 + y_3}{2};$$

If we make a slight modification in the ratios of these last coefficients by substituting for each the nearest multiple of 42, we have

$$A = \frac{H}{840}[42, 210, 42, 252, 42, 210, 42],$$

(the denominator remaining unchanged, since the sum of the coefficients is still 840), which reduces to

$$A = \frac{H}{20}[1, 5, 1, 6, 1, 5, 1].$$

This result is known as *Weddles' Rule* for six intervals. The value thus given to the mean ordinate is evidently a very close approximation to that resulting from Cotes' method, the difference being

$$\frac{1}{840}[y_1 + y_7 + 15(y_3 + y_5) - 6(y_2 + y_6) - 20y_4].$$

and substituting

$$A = \frac{h}{12}(5y_1 + 8y_2 - y_3).$$

173. It is, however, to be noticed, that when the ordinates are symmetrically situated with respect to the area, if n is *even*, the parabolic curve may be assumed of the $(n+1)$th degree. For example, in Art. 168, $n = 2$, but the curve was assumed of the third degree. Inasmuch as A, B, C and D cannot all be expressed in terms of y_1, y_2, and y_3, we see that a variety of parabolic curves of the third degree can be passed through the extremities of the measured ordinates, but all of these curves have the same area.*

Application to Solids.

174. If y denotes the area of the section of a solid perpendicular to the axis of x, the volume of the solid is $\int y\,dx$, and

---

* This circumstance indicates a probable advantage in making n an even number when repeated applications of the rules are made. Thus, in the case of six intervals, we can make three applications of Simpson's first rule, giving

$$A = \frac{H}{18}[1, 4, 2, 4, 2, 4, 1], \quad . \quad . \quad . \quad . \quad . \quad . \quad . \quad . \quad (1)$$

or two of Simpson's second rule, giving

$$A = \frac{H}{16}[1, 3, 3, 2, 3, 3, 1]. \quad . \quad . \quad . \quad . \quad . \quad . \quad . \quad . \quad (2)$$

In the first case, we assume the curve to consist of three arcs of the third degree, meeting at the extremities of the ordinates y_3 and y_5; but, since each of these arcs contains an undetermined constant, we can assume them to have common tangents at the points of meeting. We have therefore a *smooth*, though not a continuous curve. In the second case, we have two arcs of the third degree containing no arbitrary constants, and therefore making an angle at the extremity of y_4. It is probable, therefore, that the smooth curve of the first case will in most cases form a better approximation than the broken curve of the second case.

In confirmation of this conclusion, it will be noticed that the ratios of the coefficients in equation (1) are nearer to those of Cotes' coefficients for $n = 6$, given in the preceding foot-note, than are those in equation (2).

therefore the approximate rules deduced in the preceding articles apply to solids as well as to areas. Indeed, they may be applied to the approximate computation of any integral, by putting y equal to the coefficient of x under the integral sign.

The areas of the sections may of course be computed by the approximate rules.

Woolley's Rule.

175. When the base of the solid is rectangular, and the ordinates of the sections necessary to the application of Simpson's first rule are measured, we may, instead of applying that rule, introduce the ordinates directly into the expression for the area in the following manner.

Taking the plane of the base for the plane of xy, and its centre for the origin, let the equation of the upper surface be assumed of the form

$$z = A + Bx + Cy + Dx^2 + Exy + Fy^2 + Gx^3 + Hx^2y + Ixy^2 + Jy^3.$$

Let $2h$ and $2k$ be the dimensions of the base, and denote the measured values of z as indicated in Fig. 25. The required volume is

$$V = \int_{-h}^{h}\int_{-k}^{k} z\,dy\,dx.$$

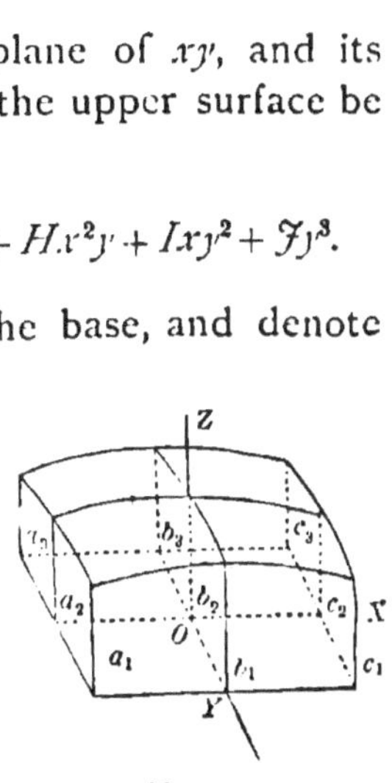

FIG. 25.

This double integral vanishes for every term containing an odd power of x or an odd power of y: hence

$$V = 4Ahk + \frac{4Dh^3k}{3} + \frac{4Fhk^3}{3},$$

$$= \frac{hk}{3}[12A + 4Dh^2 + 4Fk^2]. \quad . \quad . \quad . \quad . \quad . \quad (1)$$

By substituting the values of x and y in the equation of the surface, we readily obtain

$$b_2 = A, \quad . \quad . \quad . \quad . \quad . \quad . \quad . \quad . \quad . \quad . \quad (2)$$

$$a_1 + a_3 + c_1 + c_3 = 4A + 4Dh^2 + 4Fk^2, \quad . \quad . \quad . \quad (3)$$

$$a_2 + c_2 + b_1 + b_3 = 4A + 2Dh^2 + 2Fk^2. \quad . \quad . \quad . \quad (4)$$

From these equations two very simple expressions for the volume may be derived; for, employing (2) and (4), equation (1) becomes

$$V = \frac{2hk}{3}(a_2 + b_1 + 2b_2 + b_3 + c_2); \quad . \quad . \quad . \quad . \quad (4)$$

and employing (2) and (3),

$$V = \frac{hk}{3}(a_1 + a_3 + 8b_2 + c_1 + c_3). \quad . \quad . \quad . \quad . \quad (5)$$

Equation (4) is known as *Woolley's Rule;* the ordinates employed are those at the middles of the sides and at the centre; in (5), they are at the corners and at the centre.

Examples XIII.

1. Apply Simpson's Rule to the sphere, the hemisphere, and the cone, and explain why the results are perfectly accurate.

2. Apply Simpson's Second Rule to the larger segment of a sphere made by a plane bisecting at right angles a radius of the sphere.

$$\frac{9\pi a^3}{8}.$$

3. Find by Simpson's Rule the volume of a segment of a sphere, b and c being the radii of the bases, and h the altitude.

$$\frac{\pi h}{6}(3b^2 + 3c^2 + h^2).$$

4. Find by Simpson's Rule the volume of the frustum of a cone, b and c being the radii of the bases, and h the altitude.

$$\frac{\pi h}{3}(b^2 + bc + c^2).$$

5. Compute by Simpson's First and Second Rules, the value of $\int_0^1 \frac{dx}{1+x}$, the common interval being $\frac{1}{12}$ in each case.

The first rule gives 0.6931487, and the second rule gives 0.6931505. The correct value is obviously $\log_e 2 = 0.6931472$.

6. Find the volume considered in Art. 175, directly by Simpson's Rule, and show that the result is consistent with equations (4) and (5).

$$V = \frac{hk}{9}[a_1 + a_3 + c_1 + c_3 + 4(a_2 + b_1 + b_3 + c_2) + 16b_2].$$

7. Find, by elimination, from equations (4) and (5), Art. 175, a formula which can be used when the centre ordinate is unknown.

$$V = \frac{hk}{3}[4(a_2 + b_1 + b_3 + c_2) - (a_1 + a_3 + c_1 + c_3)].$$

CHAPTER IV.

Mechanical Applications.

XIV.

Definitions.

176. We shall give in this chapter a few of the applications of the Integral Calculus to mechanical questions.

The *mass* or quantity of matter contained in a body is proportional to its weight. When the masses of all parts of equal volume are equal, the body is said to be *homogeneous.* The factor by which it is necessary to multiply the unit of volume to produce the unit of mass is called the *density*, and usually denoted by γ.

In the following articles it will be assumed, when not otherwise stated, that the body is homogeneous, and that the density is equal to unity, so that the unit of mass is identical with the unit of volume. When the mass of an area is spoken of, it is regarded as a lamina of uniform thickness and density, and the unit of mass is taken to correspond with the unit of surface. In like manner the unit of mass for a line is taken as identical with the unit of length.

Statical Moments.

177. The *moment* of a force, with reference to a point, is the measure of the effectiveness of the force in producing motion about the point. It is shown in treatises on Mechanics, that this is *the product of the force and the perpendicular from the point upon the line of application of the force.*

The moment of the sum of a number of forces about a given point is the sum of the moments of the forces.

The *statical moment* of a body about a given point is the moment of its gravity; the force of gravity being supposed to act upon every part of the body, and in parallel lines.

178. In order to find the statical moment of a continuous body, we regard the body as generated geometrically in some convenient manner, and determine the corresponding differential of the moment.

In the case of a plane area, let the body be referred to rectangular axes, and let gravity be supposed to act in the direction of the axis of y. Then the abscissa of the point of application is the *arm* of the force when we consider the moment about the origin. Let us first suppose the area to be generated by the motion of the ordinate y. The differential of the area is then $y\,dx$. The corresponding element of the sum, of which the integral $\int_a^b y\,dx$ is the limiting value, see Art. 99, is

$$y_r \Delta x, \quad . \quad . \quad . \quad . \quad . \quad . \quad . \quad . \quad (1)$$

in which y_r is the ordinate corresponding to *any* value of x intermediate between $a + (r-1)\Delta x$, and $a + r\Delta x$. It is evident that the arm of the weight of the element (1) is such an intermediate value of x; hence the moment of the element is

$$x_r y_r \Delta x. \quad . \quad . \quad . \quad . \quad . \quad . \quad . \quad . \quad (2)$$

The whole moment is therefore the limiting value of a sum of the form

$$\Sigma_a^b x_r y_r \Delta x.$$

In other words, it is the integral

$$\int_a^b xy\,dx, \quad . \quad . \quad . \quad . \quad . \quad . \quad . \quad . \quad (3)$$

in which the differential of the moment is the product of the differential of the area and the arm of the force, which in this case is the same for every point of the element. In other words, *the moment of the differential is the differential of the moment.*

179. As an illustration, we find the moment of a semicircle (Fig. 26) about its centre. The area may be generated by the line $2y$, moving from $x = 0$ to $x = a$. The equation of the circle being

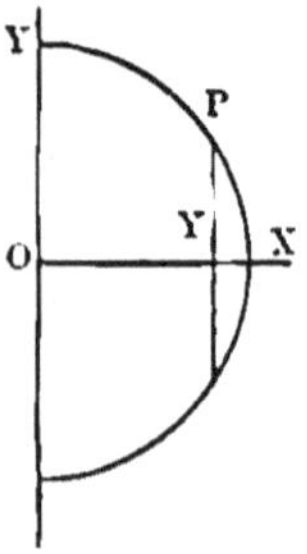

FIG. 26.

$$x^2 + y^2 = a^2,$$

the differential of the area is

$$2\sqrt{(a^2 - x^2)}\,dx.$$

The moment of this differential is

$$2\sqrt{(a^2 - x^2)}x\,dx;$$

hence the whole moment is

$$2\int_0^a \sqrt{(a^2 - x^2)}x\,dx = -\frac{2}{3}(a^2 - x^2)^{\frac{3}{2}}\Big]_0^a = \frac{2a^3}{3}.$$

Centres of Gravity.

180. If a force equal to the whole weight of a body be applied with an arm properly determined, its moment may be made equivalent to the whole statical moment of the body. If the force is in the direction of the axis of y, as in Fig. 26, we have, denoting this arm by $\bar{x}$,

$$\bar{x} \cdot \text{Area} = \text{Moment},$$

$$\bar{x} = \frac{\text{Moment}}{\text{Area}}.$$

In like manner, supposing the force to act in the direction of the axis of x, we may determine $\bar{y}$ for the same body.

It is shown in treatises on Mechanics that the point determined by the two coordinates $\bar{x}$ and $\bar{y}$, is independent of the position of the coordinate axis. This point is called the *centre of gravity* of the area. The centre of gravity of a volume is defined in like manner.

181. The symmetry of the form of a body may determine one or more of the coordinates of its centre of gravity. Thus the centre of gravity of a circle or a sphere coincides with the geometrical centre, and the centre of gravity of a solid of revolution is on the axis of revolution. The centre of gravity of the semicircle in Fig. 26, is on the axis of x; hence to determine its position we have only to find $\bar{x}$. Dividing the moment of the semicircle found in Art. 179 by the area $\frac{1}{2}\pi a^2$, we have

$$\bar{x} = \frac{4a}{3\pi}.$$

182. In finding the moment of the semicircle (Art. 179), we regarded the area as generated by the double ordinate $2y$, and the differential of the moment was found by multiplying the differential of the area by x, which is the arm of the force for every point of the generating line.

We may, however, derive the moment from the differential of area,

$$x\,dy, \quad \ldots \ldots \ldots \ldots \quad (1)$$

since the area may be generated by the motion of the abscissa x from $y = -a$ to $y = a$. But in this case to find the moment of the differential we must multiply it by the distance of its centre of gravity from the given axis. The centre of gravity of the line x is evidently its middle point, hence the required arm is $\frac{1}{2}x$. Therefore the differential of the moment is

$$\frac{x^2\,dy}{2}; \quad \ldots \ldots \ldots \ldots \quad (2)$$

and consequently the whole moment is

$$\frac{1}{2}\int_{-a}^{a} x^2\,dy = \frac{1}{2}\int_{-a}^{a}(a^2 - y^2)\,dy = \frac{2a^3}{3}.$$

This result is identical with that derived in Art. 179.

Polar Formulas.

183. When polar formulas are employed, r and θ being coordinates of the curved boundary of the area, the element is $\frac{1}{2}r^2\,d\theta$. Since this element is ultimately a triangle, we employ the well known property of triangles; that the centre of gravity is on a medial line at two-thirds the distance from the vertex to the base.

The coordinates of the centre of gravity of the element are, therefore,

$$\frac{2}{3}r\sin\theta \qquad \text{and} \qquad \frac{2}{3}r\cos\theta.$$

Hence we have the formula

$$\bar{x} = \frac{\int \frac{2}{3}r\cos\theta\,\frac{1}{2}r^2\,d\theta}{\frac{1}{2}\int r^2\,d\theta} = \frac{2}{3}\cdot\frac{\int r^3\cos\theta\,d\theta}{\int r^2\,d\theta},$$

and similarly $$\bar{y} = \frac{2}{3}\cdot\frac{\int r^3\sin\theta\,d\theta}{\int r^2\,d\theta}.$$

184. To illustrate, let us find the centre of gravity of the area enclosed by the lemniscata

$$r^2 = a^2 \cos 2\theta.$$

Whence $$\bar{x} = \frac{2a}{3} \frac{\int_0^{\frac{\pi}{4}} (\cos 2\theta)^{\frac{3}{2}} \cos\theta \, d\theta}{\int_0^{\frac{\pi}{4}} \cos 2\theta \, d\theta} = \frac{4a}{3} \int_0^{\frac{\pi}{4}} (\cos 2\theta)^{\frac{3}{2}} \cos\theta \, d\theta.$$

Put $\cos 2\theta = \cos^2\phi,$ whence $\sin\phi = \sqrt{2}\sin\theta,$

and $$\sqrt{2}\cos\theta \, d\theta = \cos\phi \, d\phi,$$

$$\therefore \quad \bar{x} = \frac{2\sqrt{2}}{3} a \int_0^{\frac{\pi}{2}} \cos^4\phi \, d\phi = \frac{2\sqrt{2}}{3} \cdot \frac{3 \cdot 1}{4 \cdot 2} \cdot \frac{\pi}{2} a = \frac{\sqrt{2}}{8} \pi a.$$

Solids of Revolution.

185. To find the centre of gravity of a solid of revolution, we take the axis of revolution as the axis of x, and the circle whose area is πy^2 as the generating element. Replacing y in equation (3), Art. 178, by this expression, we have for the statical moment

$$\pi \int_a^b xy^2 \, dx,$$

and for the abscissa of the centre of gravity

$$\bar{x} = \frac{\int_a^b xy^2 \, dx}{\int_a^b y^2 \, dx}$$

186. To illustrate, we find the centre of gravity of a spherical segment whose height is h. In this case, taking the origin at the vertex of the segment, and denoting the radius of the sphere by a, we have

$$y^2 = 2ax - x^2.$$

Hence
$$\bar{x} = \frac{\int_0^h (2ax^2 - x^3)\,dx}{\int_0^h (2ax - x^2)\,dx} = \frac{\left[\frac{2}{3}ax^3 - \frac{1}{4}x^4\right]_0^h}{\left[ax^2 - \frac{1}{3}x^3\right]_0^h} = \frac{h}{4}\cdot\frac{8a - 3h}{3a - h}.$$

If the centre of gravity of the surface of the segment be required, since the differential of the surface is $2\pi y\,ds$, we easily obtain the general formula

$$x = \frac{\int_0^h xy\,ds}{\int_0^h y\,ds},$$

and, in this case the curve being a circle, $y\,ds = a\,dx$; hence, substituting, we have

$$\bar{x} = \tfrac{1}{2}h.$$

The Properties of Pappus.

187. Let a solid be generated by the revolution of any plane figure about an exterior axis in its own plane. It is required to determine the volume and the surface thus generated.

It is evident that this solid may also be generated by a variable circular ring whose centre moves along the axis of revolution; denoting by y_1 and y_2 corresponding ordinates of the outer and inner circles respectively, the area of this ring is $\pi(y_1^2 - y_2^2)$. Hence

$$V = \pi\int(y_1^2 - y_2^2)\,dx = 2\pi\int\frac{y_1 + y_2}{2}(y_1 - y_2)\,dx.$$

But this integral is the statical moment of the given figure, since $y_1 - y_2$ is the generating element of its area, and $\frac{y_1 + y_2}{2}$ is the corresponding arm. Denoting the area of the figure by A, we may therefore write

$$V = 2\pi\bar{y}A\,;$$

that is, *the volume is the product of the area of the figure and the path described by its centre of gravity.*

The surface (S) of this solid is, by Art. 149,

$$S = 2\pi\int y\,ds = 2\pi\,\bar{y}\int ds,$$

if $\bar{y}$ denotes the ordinate of the centre of gravity of the arc s.

Hence we have $$S = 2\pi\bar{y}\cdot arc\,;$$

that is, *the surface is the product of the length of the arc into the path described by the centre of gravity.*

These theorems are frequently called the properties of Guldinus; they are, however, due to Pappus, who published them 1588.

It is obvious that both theorems are true for any part of a revolution of the generating figure.

Examples XIV.

1. Find the centre of gravity of the area enclosed between the parabola $y^2 = 4mx$ and the double ordinate corresponding to the abscissa a.

$$\bar{x} = \frac{3a}{5}.$$

2. Find the centre of gravity of the area between the semi-cubical parabola $ay^2 = x^3$ and the double ordinate which corresponds to the abscissa a.

$$\bar{x} = \frac{5a}{7}.$$

3. Find the ordinate of the centre of gravity of the area between the axis of x and the sinusoid $y = \sin x$, the limits being $x = 0$ and $x = \pi$.

$$\bar{y} = \tfrac{1}{8}\pi.$$

4. Find the coordinates of the centre of gravity of the area between the axes and the parabola

$$\left(\frac{x}{a}\right)^{\frac{1}{2}} + \left(\frac{y}{b}\right)^{\frac{1}{2}} = 1.$$

$$\bar{x} = \frac{a}{5}, \text{ and } \bar{y} = \frac{b}{5}.$$

5. Find the centre of gravity of the area between the *cissoid* $y^2(a - x) = x^3$ and its asymptote.

Solution :—

Denoting the statical moment by M and the area by A,

$$M = \int_0^a \frac{x^{\frac{5}{2}}\,dx}{(a - x)^{\frac{1}{2}}} = -2x^{\frac{5}{2}}(a - x)^{\frac{1}{2}}\Big]_0^a + 5\int_0^a x^{\frac{3}{2}}(a - x)^{\frac{1}{2}}\,dx$$

$$= 5a \cdot A - 5M;$$

$$\therefore M = \frac{5a}{6}A, \qquad \text{hence} \qquad \bar{x} = \frac{5a}{6}.$$

6. Find the centre of gravity of the area between the parabola $y^2 = 4ax$ and the straight line $y = mx$.

$$\bar{x} = \frac{8a}{5m^2}, \text{ and } \bar{y} = \frac{2a}{m}.$$

7. Find the centre of gravity of the segment of an ellipse cut off by a quadrantal chord.

$$\bar{x} = \frac{2}{3} \cdot \frac{a}{\pi - 2}, \text{ and } \bar{y} = \frac{2}{3} \cdot \frac{b}{\pi - 2}.$$

8. Given the cycloid,

$$y = a(1 - \cos\psi), \qquad x = a(\psi - \sin\psi),$$

find the distance of its centre of gravity from the base.

$$\bar{y} = \frac{5a}{6}.$$

9. Find the centre of gravity of the area enclosed between the positive directions of the coordinate axes and the four-cusped hypocycloid

$$x^{\frac{2}{3}} + y^{\frac{2}{3}} = a^{\frac{2}{3}}.$$

Put $x = a\cos^3\theta$, *and* $y = a\sin^3\theta$.

$$\bar{x} = \bar{y} = \frac{256a}{315\pi}.$$

10. Find the centre of gravity of the area enclosed by the *cardioid*

$$r = a(1 - \cos\theta).$$

$$\bar{x} = -\frac{5a}{6}.$$

11. Find the centre of gravity of the sector of a circle whose radius is a, the angle of the sector being 2α.

Use the method of Art. 183.

$$\bar{x} = \frac{2}{3}\frac{a\sin\alpha}{\alpha}.$$

12. Find the centre of gravity of the segment of a circle, the angle subtended being 2α and the radius of the circle a.

Solution :—

$$\bar{x} = \frac{2\int_{a\cos\alpha}^{a}(a^2 - x^2)^{\frac{1}{2}}x\,dx}{\text{Area}} = \frac{2a^3\sin^3\alpha}{3\ \text{Area}} = \frac{\text{Chord}^3}{12\ \text{Area}}.$$

13. Find the centre of gravity of a circular ring, the radii being a and a_1, and the angle subtended 2α.

$$\bar{x} = \frac{2}{3}\cdot\frac{a^3 - a_1^3}{a^2 - a_1^2}\cdot\frac{\sin\alpha}{\alpha}.$$

14. Find the centre of gravity of a circular arc, whose length is $2s$.

Solution :—

We have in this case, taking the origin at the centre and the axis of x bisecting the arc,

$$\bar{x} = \frac{\int_{-s}^{s} x\,ds}{\int_{-s}^{s} ds}.$$

Put $x = a\cos\theta$, then $ds = a\,d\theta$, and denoting by α the angle subtended by s, we have

$$\bar{x} = \frac{a^2\int_{-\alpha}^{\alpha}\cos\theta\,d\theta}{2s} = \frac{a\sin\alpha}{\alpha} = \frac{c}{\alpha},$$

$2c$ being the chord.

15. Find the coordinates of the centre of gravity of arc of the semi-cycloid whose equations, referred to the vertex, are

$$x = a(1 - \cos\psi), \qquad \text{and} \qquad y = a(\psi + \sin\psi).$$

$$\bar{x} = \frac{2a}{3}, \text{ and } \bar{y} = \left(\pi - \frac{4}{3}\right) a.$$

16. Find the centre of gravity of the arc between two successive cusps of the four-cusped hypocycloid

$$x^{\frac{2}{3}} + y^{\frac{2}{3}} = a^{\frac{2}{3}}.$$

$$\bar{x} = \bar{y} = \frac{2a}{5}.$$

17. Find the position of the centre of gravity of the arc of the semi-cardioid

$$r = a(1 - \cos\theta).$$

$$\bar{x} = -\frac{4a}{5}, \text{ and } \bar{y} = \frac{4a}{5}.$$

18. A semi-ellipsoid is formed by the revolution of a semi-ellipse about its major axis; find the distance of the centre of gravity of the solid from the centre of the ellipse.

$$\bar{x} = \frac{3a}{8}.$$

19. Find the centre of gravity of a frustum of a paraboloid of revolution having a single base, h denoting the height of the frustum.

$$\bar{x} = \frac{2h}{3}.$$

20. A paraboloid and a cone have a common base and vertices at the same point; find the centre of gravity of the solid enclosed between them.

The centre of gravity is the middle point of the axis.

21. Find the centre of gravity of a hyperboloid whose height is h, the generating curve being

$$y^2 = m(2ax + x^2).$$

$$\bar{x} = \frac{h}{4} \cdot \frac{8a + 3h}{3a + h}.$$

22. Find the centre of gravity of the solid formed by the revolution of the sector of a circle about one of its extreme radii.

The height of the cone being denoted by h, and the radius of the circle by a, we have

$$\bar{x} = \frac{3}{8}(a + h).$$

23. Find the centre of gravity of the solid formed by the revolution about the axis of x of the curve

$$a^2 y = ax^2 - x^3,$$

between the limits 0 and a.

$$\bar{x} = \frac{5a}{8}.$$

24. A solid is formed by revolving about its axis the cardioid

$$r = a(1 - \cos\theta);$$

find the distance of the cusp from the centre of gravity.

$$\bar{x} = \frac{16a}{15}.$$

25. Determine the position of the centre of gravity of the volume included between the surfaces generated by revolving about the axis of x the two parabolas

$$y^2 = mx, \qquad \text{and} \qquad y^2 = m'(a - x).$$

$$\bar{x} = \frac{a}{3} \cdot \frac{m + 2m'}{m + m'}.$$

26. Find the centre of gravity of a rifle bullet consisting of a cylinder two calibers in length, and a paraboloid one and a half calibers in length having a common base, the opposite end of the cylinder containing a conical cavity one caliber in depth with a base equal in size to that of the cylinder.

The distance of the centre of gravity from the base of the bullet is $1\frac{33}{58}$ calibers.

27. A solid formed by the revolution of a circular segment about its chord is cut in halves by a plane perpendicular to the chord; determine the centre of gravity of one of the halves. This solid is called an *ogival.*

Denoting by 2α the angle subtended by the chord, and by a the radius of the circle, the distance of the centre of gravity from the base is

$$\bar{x} = \frac{a}{16} \cdot \frac{44 \sin^2 \alpha + \sin^2 2\alpha + 32 (\cos 2\alpha - \cos \alpha)}{\sin \alpha (2 + \cos^2 \alpha) - 3\alpha \cos \alpha}.$$

28. Find the centre of gravity of the surface of the paraboloid formed by the revolution about the axis of x of the parabola

$$y^2 = 4mx,$$

a denoting the height of the paraboloid.

$$\bar{x} = \frac{1}{5} \cdot \frac{(3a - 2m)(a + m)^{\frac{3}{2}} + 2m^{\frac{5}{2}}}{(a + m)^{\frac{3}{2}} - m^{\frac{3}{2}}}.$$

29. Find the centre of gravity of the surface generated by the revolution of a semi-cycloid about its axis, the equations of the curve being

$$x = a(1 - \cos \psi), \qquad \text{and} \qquad y = a(\psi + \sin \psi).$$

$$\bar{x} = \frac{2a}{15} \cdot \frac{15\pi - 8}{3\pi - 4}.$$

30. Find the centre of gravity of the surface generated by the revolution about its axis of one of the loops of the lemniscata

$$r^2 = a^2 \cos 2\theta.$$

$$\bar{x} = \frac{2 + \sqrt{2}}{6} a.$$

31. A cardioid revolves about its axis; find the centre of gravity of the surface generated, the equation of the cardioid being

$$r = a(1 - \cos\theta).$$

$$\bar{x} = \frac{50a}{63}.$$

32. A ring is generated by the revolution of a circle about an axis in its own plane; c being the distance of the centre of the circle from the axis, and a the radius, determine the volume and surface generated.

$$V = 2\pi^2 ca^2, \text{ and } S = 4\pi^2 ca.$$

33. A triangle revolves about an axis in its plane; a_1, a_2, and a_3, denoting the distances of its vertices from the axis, determine the volume generated.

$$V = \frac{2\pi A}{3}(a_1 + a_2 + a_3).$$

34. Find the volume of a frustum of a cone, the radii of the bases being a_1 and a_2, and the height h.

$$V = \frac{\pi h}{3}(a_1^2 + a_1 a_2 + a_2^2).$$

35. Find the volume and surface generated by the revolution of a cycloid about its base.

$$V = 5\pi^2 a^3, \text{ and } S = \frac{64\pi a^2}{3}.$$

XV.

Moments of Inertia.

188. When a body rotates about a fixed axis, the velocity of a particle at a distance r from the axis is

$$r\frac{d\omega}{dt},$$

in which ω is the angle of rotation. The force which acting for a unit of time would produce this motion in a mass m is measured by the momentum

$$mr\frac{d\omega}{dt}.$$

The moment of this force about the axis is therefore

$$mr^2\frac{d\omega}{dt}.$$

The sum of these moments for all the parts of a rigid system is

$$\frac{d\omega}{dt}\Sigma\Delta mr^2,$$

since the angular velocity, $\frac{d\omega}{dt}$, is constant. In the case of a continuous body this expression becomes

$$\frac{d\omega}{dt}\int r^2\,dm,$$

in which dm is the differential of the mass. The factor

$$\int r^2\,dm,$$

which depends upon the shape of the body, is called its *moment of inertia*, and is denoted by I.

189. When the body is homogeneous, dm is to be taken equal to the differential of the line, area, or volume, as the case may be. For example, in finding the moment of inertia of a straight line whose length is $2a$, about an axis bisecting it at right angles, we let x denote the distance of any point from the axis; then $dm = dx$, hence we have

$$I = \int_{-a}^{a} x^2\,dx = \frac{2a^3}{3} = \frac{(2a)^3}{12}.$$

Again, in finding the moment of inertia of the semi-circle in figure 25, about the axis of y, let $dm = 2y\,dx$; then, since every point of the generating line is at the distance x from the axis, the moment of inertia is

$$I = 2\int_0^a yx^2\,dx = 2\int_0^a \sqrt{(a^2 - x^2)}\,x^2\,dx.$$

Putting $x = a\sin\theta$, we have

$$I = 2a^4\int_0^{\frac{1}{2}\pi} \cos^2\theta \sin^2\theta\,d\theta = \frac{\pi a^4}{8}.$$

The Radius of Gyration.

190. If the whole mass of the body were situated at the distance k from the axis, its moment of inertia would be k^2m. Now, if k is so determined that *this moment* shall be equal to the actual moment of inertia of the body, the value of k is *the radius of gyration* of the body with reference to the given axis. Hence

$$k^2 = \frac{\text{Moment of inertia}}{\text{Mass}}.$$

Thus, for the radius of gyration of the line $2a$, whose moment of inertia is found in the preceding article, we have

$$k^2 = \frac{a^2}{3}, \qquad \text{or} \qquad k = \frac{a}{\sqrt{3}};$$

and for the radius of gyration of the semi-circle, whose area is $\frac{1}{2}\pi a^2$,

$$k^2 = \frac{a^2}{4}, \qquad \text{or} \qquad k = \frac{a}{2}.$$

It is evident that this expression is also the radius of gyration of the whole circle about a diameter, for the moment of inertia of the circle is evidently double that of the semi-circle, and its area is also double that of the semi-circle.

191. It is sometimes convenient to use modes of generating the area or volume, other than those involving rectangular coordinates. For example, let it be required to find the radius of gyration of a circle whose radius is a, about an axis passing through its centre and perpendicular to its plane. This circle may be generated by the circumference of a variable circle whose radius is r, while r passes from 0 to a. The differential of the area is then $2\pi r\,dr$, and the moment is

$$I = 2\pi \int_0^a r^3\,dr = \frac{\pi a^4}{2}.$$

Dividing by the area of the circle, we have

$$k^2 = \frac{a^2}{2}.$$

192. Again, to find the radius of gyration of a sphere whose radius is a about a diameter. In order that all points of the elements shall be at the same distance from the axis we regard the sphere as generated by the surface of a cylinder whose radius is x, and whose altitude is $2y$. The surface of this cylinder is therefore $4\pi xy$. The differential of the volume is $4\pi xy\,dx$, and the moment of inertia is

$$I = 4\pi \int x^3 y\,dx = 4\pi \int \sqrt{(a^2 - x)}\,x^3\,dx.$$

Putting $x = a \sin \theta$,

$$I = 4\pi a^5 \int_0^{\frac{\pi}{2}} \sin^3 \theta \cos^2 \theta\,d\theta = \frac{8\pi a^5}{15}.$$

Dividing by $\frac{4\pi a^3}{3}$, the volume of the sphere, we have

$$k^2 = \frac{2a^2}{5}.$$

Radii of Gyration about Parallel Axes.

193. *The moment of inertia of a body about any axis exceeds its moment of inertia about a parallel axis passing through the centre of gravity, by the product of the mass and the square of the distance between the axes.*

Let h be the distance between the axes. Pass a plane through the element dm perpendicular to the axes, and let r and r_1 be the distances of the element from the axes. Then, r, r_1, and h form a triangle; let θ be the angle at the axis passing through the centre of gravity, then

$$r^2 = r_1^2 + h^2 - 2r_1 h \cos \theta. \quad . \quad . \quad . \quad . \quad . \quad (1)$$

The moment of inertia is therefore

$$\int r^2 dm = \int r_1^2\, dm + h^2 m - 2h \int r_1 \cos\theta\, dm \quad . \quad . \quad . \quad (2)$$

Now r_1 and θ are the polar coordinates of dm, in the plane which is passed through the element; hence the last integral in equation (2) is equivalent to

$$-2h \int x\, dm.$$

But $\int x\, dm$ is the statical moment of the body about the axis passing through the centre of gravity. Now from the definition of the centre of gravity, this moment is zero; hence, equation (2) reduces to

$$\int r^2\, dm = \int r_1^2\, dm + h^2 m \quad . \quad . \quad . \quad . \quad . \quad (3$$

Introducing the radii of gyration, we have also

$$k^2 = k_1^2 + h^2. \quad . \quad . \quad . \quad . \quad . \quad . \quad . \quad . \quad (4)$$

194. As an application of this result, we shall now find the moment of inertia of a cone whose height is h, and the radius of whose base is a, about an axis passing through its vertex perpendicular to its geometrical axis. Taking the origin at the vertex of the cone, the axis of x coincident with the geometrical axis, and a circle perpendicular to this axis as the generating element, we have for the area of this element πy^2, and for its radius of gyration about a diameter parallel to the given axis, $\frac{y^2}{4}$.

The distance between these axes being x, the proposition proved in the preceding article gives an expression for the radius of gyration of the element about the given axis; viz., $x^2 + \frac{y^2}{4}$. Replacing r^2, in the general expression for I (Art. 188), by this expression, and substituting for dm the differential $\pi y^2\,dx$, we have

$$I = \pi \int \left(x^2 + \frac{y^2}{4}\right) y^2\,dx,$$

in which $y = \frac{ax}{h}$. Therefore

$$I = \frac{\pi a^2}{h^2}\int_0^h \left(1 + \frac{a^2}{4h^2}\right) x^4\,dx = \frac{\pi a^2 h^3}{5}\left(1 + \frac{a^2}{4h^2}\right),$$

and since

$$V = \frac{\pi a^2 h}{3},$$

$$k^2 = \frac{3}{20}\left(a^2 + 4h^2\right).$$

To find the square of the radius of gyration about a parallel axis through the centre of gravity, we have

$$k_0^2 = \frac{3}{20}\left(a^2 + 4h^2\right) - \left(\frac{3h}{4}\right)^2$$

$$= \frac{3}{80}\left(4a^2 + h^2\right).$$

To find the moment of inertia of a right cone about its geometrical axis we employ the same generating element as before; but in this case the square of the radius of gyration is $\frac{y^2}{2}$. Hence

$$I = \frac{\pi}{2}\int y^4\,dx = \frac{\pi a^4}{2h^4}\int_0^h x^4\,dx;$$

therefore

$$I = \frac{\pi a^4 h}{10}, \quad \text{whence} \quad k^2 = \frac{3a^2}{10}.$$

Polar Moments of Inertia.

195. In the case of a plane area, when the axis of rotation passes through the origin, we have

$$r^2 = x^2 + y^2, \qquad \text{hence} \int r^2\,dm = \int (x^2 + y^2)\,dm,$$

therefore

$$I = \int x^2\,dm + \int y^2\,dm;$$

that is, *the sum of the moments of inertia of a plane area about two axes in its own plane at right angles to each other is equal to the moment of inertia about an axis through the origin perpendicular to the plane.* I in the above equation is called *the polar moment of inertia.*

In the case of the circle, since the moment is the same about every diameter, the polar moment is twice the moment about a diameter; that is, denoting the former by I_p and the latter by I_a, we have

$$I_p = 2\,I_a = \frac{\pi a^4}{2}.$$

See Art. 191.

Examples XV.

1. Find the radius of gyration of a circular arc ($2s$) about a radius passing through its vertex.

Solution :—

Taking the origin at the centre, and the axis of x bisecting the arc, and denoting by 2α the angle subtended by $2s$, we have

$$mk^2 = \int_{-s}^{s} y^2\, ds = a^3 \int_{-\alpha}^{\alpha} \sin^2\theta\, d\theta.$$

$$m = 2a\alpha \qquad \therefore \qquad k^2 = \frac{a^2}{2}\left(1 - \frac{\sin 2\alpha}{2\alpha}\right)$$

2. Find the radius of gyration of the same arc about the axis of y, and thence about a perpendicular axis through the centre of the circle. $k = a$.

3. Find the radius of gyration of the same arc about an axis through its vertex perpendicular to the plane of the circle.
See Ex. XIV., 14, *and denote by* c *the subtending chord.*

$$k^2 = 2a^2\left(1 - \frac{c}{2s}\right).$$

4. Find the moment of inertia of the chord of a circular arc, in terms of the diameter parallel to it, and its angular distance from this diameter.

See Arts. 189 *and* 193. $I = \dfrac{d^3}{24}(3\cos\alpha - \cos 3\alpha)$.

5. Find the radius of gyration of an ellipse about an axis through its centre perpendicular to its plane.
Find the radius of gyration about the major axis and about the minor axis, and apply Art. 195.

$$k^2 = \tfrac{1}{4}(a^2 + b^2).$$

6. Find the radius of gyration of an isosceles triangle about a perpendicular let fall from its vertex upon the base ($2b$).

$$k^2 = \frac{b^2}{6}.$$

7. Find the radius of gyration about the axis of the curve, of the area enclosed by the two loops of the lemniscata

$$r^2 = a^2 \cos 2\theta.$$

$$k^2 = \frac{a^2}{48}(3\pi - 8).$$

8. Find the radius of gyration of a right triangle, whose sides are a and b. about an axis through its centre of gravity perpendicular to its plane.

$$k^2 = \frac{a^2 + b^2}{18}.$$

9. Find the radius of gyration of a portion of a parabola bounded. by a double ordinate perpendicular to the axis, about a perpendicular to its plane passing through its vertex.

$$k^2 = \tfrac{3}{7} x^2 + \tfrac{1}{5} y^2.$$

10. Find the radius of gyration of a cylinder about a perpendicular that bisects its geometrical axis, $2l$ being the length of the cylinder, and a the radius of its base.

$$k^2 = \frac{a^2}{4} + \frac{l^2}{3}.$$

11. Find the radius of gyration of a concentric spherical shell about a tangent to the external sphere, the radii being a and b.

$$k^2 = \frac{7a^5 - 5a^2b^3 - 2b^5}{5(a^3 - b^3)}.$$

12. Find the radius of gyration of a paraboloid of revolution about its axis, in terms of the radius (b) of the base.

$$k^2 = \frac{b^2}{3}.$$

13. Find the moment of inertia of an ellipsoid about one of its principal axes.

$$I = \frac{4\pi abc}{15}(b^2 + c^2).$$

14. Find the radius of gyration of a symmetrical double convex lens about its axis, a being the radius of the circular intersection of the two surfaces, and b the semi-axis.

$$k^2 = \frac{b^4 + 5a^2b^2 + 10a^4}{10(b^2 + 3a^2)}.$$

15. Find the radius of gyration of the same lens about a diameter to the circle in which the spherical surfaces intersect.

$$k^2 = \frac{10a^4 + 15a^2b^2 + 7b^4}{20(b^2 + 3a^2)}.$$

THE END.

www.ingramcontent.com/pod-product-compliance
Lightning Source LLC
Chambersburg PA
CBHW021527210326
41599CB00012B/1412
9783337308544